The Chemisorptive Bond

BASIC CONCEPTS

This is Volume 32 of
PHYSICAL CHEMISTRY
A series of monographs
Edited by ERNEST M. LOEBL, *Polytechnic Institute of Brooklyn*

A complete list of the books in this series appears at the end of the volume.

The Chemisorptive Bond

BASIC CONCEPTS

ALFRED CLARK

School of Chemical Engineering
and Materials Science
The University of Oklahoma
Norman, Oklahoma

ACADEMIC PRESS New York and London 1974

A Subsidiary of Harcourt Brace Jovanovich, Publishers

ACADEMIC PRESS, INC.
111 Fifth Avenue, New York, New York 10003

United Kingdom Edition published by
ACADEMIC PRESS, INC. (LONDON) LTD.
24/28 Oval Road, London NW1

Library of Congress Cataloging in Publication Data

Clark, Alfred, Date
The chemisorptive bond: basic concepts.

(Physical chemistry, a series of monographs, no. 32)
Includes bibliographical references.
1. Chemisorption. 2. Chemical bonds. I. Title. II. Series.
QD547.C55 547'.1'224 73-9421
ISBN 0-12-175440-5

PRINTED IN THE UNITED STATES OF AMERICA

Contents

Preface

This book describes the basic concepts of the chemisorptive bond on solid surfaces from the simple analogies with ordinary chemical bonds to the recent quantum-mechanical approaches that accept the difficult challenge of the solid state and the presence of a surface. The latter are usually omitted because of their mathematical complexity. Here, the blow of mathematics has been softened drastically and the physical pictures stressed.

The earlier chapters are obviously relevant to chemisorption, and so to catalysis, because they present simple formulas for correlating measurable quantities. Unfortunately, these simple correlations, faithful or not, often fail to give true pictures of the mechanisms of chemisorption. Although the later chapters provide more detailed, quantum-mechanical pictures, many of them cannot be checked experimentally, because techniques have not been developed yet for measuring the variables involved. Yet, these chapters should not be considered irrelevant to chemisorption phenomena. Many predictions, long resistant to experimental confirmation, have ultimately found a place in the bedrock of science: Dalton's atomic hypothesis, for example, and, much later, nuclear fission. We believe the concepts reviewed here are important and relevant. Some may wither under the scrutiny of future experiments; others may bloom. We hope none will be sentenced to death under a myopic edict of immediate relevancy.

The author wishes to thank Mrs. Sue Ryan for skillfully and patiently typing the manuscript; the University of Oklahoma, Norman, for providing the facilities used in preparing the manuscript and figures; and Academic Press for their excellent editorial and stylistic suggestions.

I

Introduction

1.1 The Forces of Adsorption

The cohesive forces in solids and the forces of adsorption are not different in principle from the forces exerted between free atoms and molecules. *All* interparticle forces have a common origin in the electromagnetic interactions of nuclei and electrons. Some of these forces are weak, like the van der Waals forces, which are considered to be made up of London dispersion forces and classical electrostatic forces. Neither transfer nor sharing of electrons occurs. Electrons may seek a new equilibrium distribution, but they always stay with the particle on which they entered the interaction. When an atom or molecule is bound to the surface of a solid by van der Waals forces, the phenomenon is called *physical adsorption.* Other forces, usually stronger, form chemical bonds and involve the transfer or sharing of electrons. When an atom or molecule is bound to a surface through overlapping of one or more of its electron orbitals, the phenomenon is called *chemisorption.* In physical adsorption, it is conventional to assume that the surface of the adsorbent is unperturbed, that its only function is to supply a potential field for the adsorbate, and that the adsorbate may be regarded as a separate thermodynamic phase. In chemisorption, these are not fruitful assumptions. The surface and the adsorbate should be viewed together as a new chemical entity. How many surface atoms of the adsorbent participate in the bonding of a single adsorbed particle is a question that has not been answered unequivocally for any real system and probably will not be in the near future. But theoretical and experimental attacks are making important advances. Chemisorptive bonds and conventional bonds may arise from the same electromagnetic forces, but in the surface bond these forces are far more intricately compounded. Atoms or ions in the surface of a solid must differ in some degree from their simpler, free counterparts because of the cooperative forces which bind them together, and often these forces create local perturbations so drastic that they have no free counterparts at all, such as interstitial atoms or ions and lattice vacancies. The chemisorption of a foreign atom on the perturbed centers of a solid surface, singly or collectively, presents grave, new problems, not fully encountered in the conventional bond, which demand separate treatment.

1.2 Approaches to the Chemisorptive Bond

Concepts of the chemisorptive bond have progressed in the usual manner from the simple to the complex. Early theories essentially ignore the presence of a surface and frequently presume that the adsorbing atom seeks out and reacts with a single surface atom. Both covalent and ionic bond formation have been considered in this analogy with the diatomic molecule. Various approximations of the strength of the covalent bond between free atoms have been adapted to adsorption. For example, Eley [1] adapted Pauling's [2] approximation to the adsorption of hydrogen atoms on metals, where the metal–hydrogen bond strength is taken as the arithmetic or geometric average of the metal–metal and the hydrogen molecule bond strengths. The metal–metal bond strength is estimated from the latent heat of sublimation of the metal, taking into account the number of nearest neighbors. Corrections have been applied for polarization effects caused by the difference in electronegativities of metal and hydrogen, giving a small ionic character to the bond [3]. Formal quantum-mechanical calculations have been made for the covalent adsorption bond on the surfaces of metals [4] and heteropolar lattices [5], and even on lattice defects [6]. Many attempts have been made to correlate empty d orbitals or d-band vacancies in transition metals with the power to chemisorb, including the application of Pauling's d character of metals, which he derived from his valence-bond theory. A great number of strictly qualitative theories representing pictorially the complexing of simple gases, olefins, and other molecules on metal and metal oxide surfaces have also been put forth, using a molecular orbital or simpler design [7].

Ionic bonds on metal surfaces have been considered in analogy with the conventional ionic bond [8]. For example, the energy change associated with the ionic mechanism for adsorption of atom A on metal M,

$$M + A \rightarrow M^{-}A^{+}$$

may be estimated by breaking down the energy as follows: (i) remove an electron from the highest occupied level of an isolated atom of A to infinity $(-Ie)$; (ii) transfer the electron to the Fermi level of the metal M (ϕe); (iii) bring A^+ to its equilibrium distance z^* from the surface of the metal $(e^2/4z^*)$. The heat of adsorption per gram atom at zero coverage is then considered to be given by

$$q_d = (-Ie + \phi e + e^2/4z^*)/N$$

where I and ϕ are the ionization potential of A and the work function of M, respectively, N is Avogadro's number, and the last term to the right in the numerator is the classical electrostatic image energy of attraction of the ion A^+ to the surface of the metal. Occasionally, the mirror-image term is replaced by a strictly interatomic potential function.

Concessions to the presence of a solid surface appear in some of the approaches which we have just described; for example, in the estimation of the metal–metal bond strength from the latent heat of sublimation in Eley's adaptation to chemisorption of Pauling's method for determining covalent bond strength, in the correlation of Pauling's d character of transition metals with chemisorption, and in the use of the mirror-image force to express the strength of the ionic bond of chemisorption on metals. But in all of these approaches, the chemisorptive bond may be regarded as localized—a quasi-diatomic molecule composed of an adsorbed atom, molecule, or ion and a surface entity of the solid lattice. The analogies with conventional bonds, covalent or ionic, persist.

Modern developments in solid-state theory have sparked fresh viewpoints on chemisorption. Applications began about 1950 and have gained momentum ever since. With their growth, analogies between the chemisorptive bond and the conventional bond are fading. Yet many areas remain equivocal; for example, the empirical studies of the change in the chemisorptive and catalytic behavior of transition metals on progressive filling of d-band vacancies by metals of group IB. The studies fail to give clear evidence on the state of the chemisorptive bond, whether it is localized outside the energy limits of the d band or nonlocalized within it. In other areas, more precise models have been developed. Preeminent is the theory of ionic chemisorption on semiconductors, called the *boundary-layer* theory [9], which depends on the donation of electrons by the valence band or the conduction band to the adsorbate, or the reverse, and on the building up of a potential field in the surface layers as adsorption proceeds. The maximum amount of adsorption is governed by the availability of electrons in the valence band or the conduction band, or by their capacities to accept electrons. The energy of adsorption is determined by the surface potential field, which changes with the amount adsorbed. Therefore, both quantities, amount and energy of adsorption, are sharply controlled by cooperative phenomena which have no existence in the free state. Although the conciseness and generality of the boundary-layer theory are appealing, experiments do not always support it. Perhaps the most serious fault of the theory is that it considers only those types of adsorption that result from a transfer of charge. There are examples of systems in which adsorption on semiconductors occurs with no change in conductivity at all [10], thus, presumably with no transfer of charge.

All approaches discussed so far have at least two serious drawbacks in common. First, each theory is built around an a priori selection of the type of chemisorptive bond, covalent or ionic, localized or nonlocalized. In a more powerful theory, the nature of the bond would evolve from the application of the theory itself. Second, it is often assumed, tacitly at least, that the forces and bonds in a solid surface and those that emerge from the surface and are available for bonding foreign atoms are not essentially different from those operating within the bulk of the solid. But the surface is a discontinuity, and this fact alone should arouse speculations on the existence of states peculiar to the surface. In the light of these limitations, the search for more rewarding theories based on solid-state theories continues.

Full quantum-mechanical treatments of surface states and chemisorption, not mere adjuncts to narrowly defined systems, are now beginning to appear. Of course, it has been known in theory since the 1930s, from the work of Tamm [11] and Shockley [12], that states peculiar to a free surface can exist. And it has been suspected that these free-surface states could have a profound effect on our theories of chemisorption. But the complexity of the problem has deterred progress. The basic quantum-mechanical developments of Baldock [13] and Koster and Slater [14] provided an important spur. Equipped with these developments, a number of workers, notably Koutecky [15], Grimley [16], and Mark [17], started the attack on the problem in the 1960s. So far, the work deals chiefly with simple models, often one-dimensional, and employs the linear combination of atomic orbitals (LCAO) approximation of the molecular orbital theory. No firm numerical values of energies have been calculated yet, but the results are interesting and provocative, for they have led to general conclusions which, at least, invite experiment. Ingenious experiments for investigating surface states have already been devised by Morrison [18] and Gray and co-workers [19].

The great importance of the theoretical studies is that, for the first time, criteria for localized and nonlocalized surface states have been clarified by the applications of basic principles. These criteria emerge in spite of the general nature of the theoretical solutions, dealing with arrays, often one-dimensional for mathematical tractability, of arbitrary surface atoms. The path which must be traveled to solve more realistic models is known and it is not an easy one. Grimley [20] has speculated about hydrogen chemisorption on metals, using quantum-mechanical methods similar to those for the more general cases. Mark [21] has pushed into the area of surface states and chemisorption on ionic lattices. We shall discuss these fascinating modern approaches in later chapters.

The application of molecular orbital procedures just cited ignores the effects of near-neighbor atoms or ions on the nature of the d orbitals. These orbitals are split in a manner that depends on the crystallographic arrangement of the neighboring atoms or ions, and the theory which takes such splitting into account is known as *crystal-field* or *ligand-field* theory [22]. Recently, there have been qualitative and empirical attempts to apply such refinements to chemisorption [23]. It is difficult to guess where they will lead, for they have not yet passed beyond the point of initial speculation. We make no predictions beyond the crystal- or ligand-field approach to the theory of the chemisorptive bond. Yet we could travel far down this road and the *real* chemisorptive bond would remain uncomfortably distant.

1.3 Real Surfaces and Chemisorptive Bonds

Real surfaces have many complexities that will probably never get incorporated into theories, and these complexities are transmitted to the chemisorptive bond. Solid surfaces are not necessarily at equilibrium. They may be "frozen" into a nonequilibrium state, and theories of nonequilibrium states tend to be arbitrary. Whether at equilibrium or not, solid surfaces may be perturbed and irregular, unlike the uniform arrays assumed in theoretical studies. Perturbations and irregularities can take many different forms. They can lead to a spectrum of adsorption-site energies and an arbitrary geographical distribution of sites. They can lead to the presence of drastically different types of sites, such as lattice vacancies and interstitial atoms or ions. Recently, it has been shown that surfaces can be completely reconstructed by the process of chemisorption [24], especially in the chemisorption of oxygen dissociatively on metals, where metal atoms are drastically rearranged upon interaction with adsorbate, so that the surface is not a neat array of metal atoms with oxygen atoms above them, but a jumble of both of them. Finally, chemisorbed particles may interact with each other, causing obscure changes in the chemisorptive bond. Although theories may never conquer the real surface and chemisorptive bond, there is plenty of room for optimism short of this goal.

REFERENCES

1. D. D. Eley, *Discuss. Faraday Soc.* **8,** 34 (1950).
2. L. Pauling, "The Nature of the Chemical Bond." Cornell Univ. Press, Ithaca, New York, 1939.
3. M. Malone, *J. Chem. Phys.* **1,** 197 (1933).
4. S. M. Kogan and V. B. Sandomirsky, *Bull. Acad. Sci. USSR Ser. Chem.* 1681 (1959).
5. Th. Wolkenstein and V. L. Bonch-Bruevich, *J. Exptl. Theoret. Phys.* (*USSR*) **20,** 624 (1950).
6. V. L. Bonch-Bruevich, *J. Phys. Chem.* (*USSR*) **27,** 662, 960 (1953).
7. W. E. Garner and F. J. Veal, *J. Chem. Soc.* 1487 (1935); P. Cossee, *J. Catal.* **3,** 80 (1964); J. J. Rooney, *ibid.* **2,** 53 (1963).
8. R. C. L. Bosworth, *Proc. Roy. Soc.* (*London*) *Ser. A* **162,** 32 (1937).
9. K. Hauffe and H. J. Engell, *Z. Elektrochem.* **56,** 366 (1952).
10. Y. Kubokawa and O. Toyama, *J. Phys. Chem.* **60,** 833 (1956).
11. I. Tamm, *Phys. Z. Sowjet* **1,** 733 (1932).
12. W. Shockley, *Phys. Rev.* **56,** 317 (1939).
13. G. R. Baldock, *Proc. Cambridge Phil. Soc.* **48,** 457 (1952).
14. G. F. Koster and J. C. Slater, *Phys. Rev.* **95,** 1167 (1954).
15. J. Koutecky, *Advan. Chem. Phys.* **9,** 85 (1965).
16. T. B. Grimley, *Advan. Catal.* **12,** 1 (1960).
17. P. Mark, *Catal. Rev.* **1,** 165 (1967).
18. S. R. Morrison, Private communication (1966).
19. T. J. Gray and P. Amigues, *Surface Sci.* **13,** 209 (1969); T. J. Gray and R. S. Cichowski, Ph.D. Thesis, N. Y. State College of Ceram. at Alfred Univ. (1968).
20. T. B. Grimley, "Chemisorption" (W. E. Garner ed.), Butterworth, London and Washington D.C., 1957.
21. P. Mark, *J. Phys. Chem. Solids* **29,** 689 (1968).
22. L. E. Orgel, "An Introduction to Transition-Metal Chemistry." Wiley, New York, 1960.
23. G. C. Bond, *Discuss. Faraday Soc.* **41,** 200 (1966).
24. J. W. May, *Surface Sci.* **18,** 431 (1969).

II

Analogies with Simple Chemical Bonds—The Covalent Bond

The pure covalent bond and the pure ionic bond are limits that the real chemisorptive bond never reaches. In this chapter, we shall discuss those theories that assume a predominantly covalent bond and make only small corrections for ionic character. In the following chapter, we shall discuss the ionic bond lying at the opposite end of the spectrum. Intermediate bonds with significant covalent and ionic character are difficult to fit into the simple models of this chapter and the next, which make a priori selections of bond type. Further, we shall assume that these bonds are analogous to the bonds in diatomic molecules. This does not mean that concessions to the presence of a surface are not made. For example, electronegativities, from which the ionic character of a covalent bond is determined, are related to the properties of the solid or its surface. Similarly, the character of the d orbitals of transition metals, often linked in theory to chemisorption, is based on the metallic state and not on the isolated atom. Yet the chemisorptive bonds considered in this chapter and the next do preserve the analogy with the localized, simple chemical bond. Each bond type links an adsorbed atom, molecule, or ion to a surface entity of the solid lattice despite that entity's association with other surface atoms, ions, or defects.

We shall consider first Eley's extension of Pauling's rules for bond energies between single atoms to adsorption bond energies. A brief discussion of other methods of determining heats of chemisorption follows, including their correlation with heats of formation of simple compounds and an empirical equation of heats of chemisorption developed by Tanaka and Tamaru. These methods tacitly assume the positioning of the adsorbed atom directly above the surface atom with which it is interacting. Speculations concerning other possible positions for the case of the adsorbed hydrogen atom have been advanced. We shall comment on the vast number of attempts to correlate heats of adsorption with simple properties of the substrate. The continued search for a single property of metals responsible for chemisorption led to the application of Pauling's valence-bond theory and attempts to correlate d character with heats of chemisorption. Next we shall describe the formal quantum-mechanical calculations of chemisorption which have been made in the light of localized bonding, to which this chapter is restricted. Broadly, all of these approaches to the covalent chemisorptive bond are presumed to involve unused metal–metal orbitals, empty d orbitals, or surface defects. Recently, there has been a great deal of research, aided by such tools as infrared spectroscopy, electron spin resonance, and low-energy electron diffraction (LEED), directed at obtaining a physical picture of the state of a chemisorbed complex. The covalent surface complexes reviewed in this chapter assume, at least tacitly, a localized bonding.

We should not expect faithful predictions of all chemisorption systems from the simple models of this chapter. At most, they are valid only for the area of predominantly covalent chemisorptive bonds. But they do provide insight and so possess heuristic value.

2.1 Estimation of Bond Strengths

In 1932, Pauling [1] suggested that the covalent bond energy of a diatomic molecule A–B could be estimated by the arithmetic averages of the homonuclear bond energies of A–A and B–B,

$$D_{AB} = (D_{AA} + D_{BB})/2. \tag{1}$$

In order to correct for the polarity or ionic character of the bond, he added another term, involving the difference in electronegativities of atoms A and B. The complete equation is

$$D_{AB} = (D_{AB} + D_{BB})/2 + 23.06(\chi_A - \chi_B)^2 \tag{2}$$

where χ_A and χ_B are the electronegativities of A and B, respectively; $(\chi_A + \chi_B)^2$ is in electron volts and the constant converts to kilocalories per mol. Equation (2) may be used to estimate the bond energies of hydrogen halides, for example, and also the heat of the reaction $H_2 + X_2 \rightarrow 2HX$, which is given by

$$Q = 2D_{HX} - D_{HH} - D_{XX}. \tag{3}$$

For the bond energies of metal hydrides, Pauling found that better agreement with experiment was obtained if the geometrical mean, $(D_{AA}D_{BB})^{1/2}$, was used in place of the arithmetic mean.

In 1950, Eley [2] adapted Pauling's approximations of the covalent bond between free atoms to adsorption bonds. He worked particularly with hydrogen–metal systems; and for dissociative adsorption,

$$2M + H_2 \rightarrow 2(M—H) \tag{4}$$

he assumed that the heat of adsorption at zero coverage is given by

$$Q_0 = 2D_{MH} - D_{HH}. \tag{5}$$

By comparison with Eq. (3), it is clear that Eley assumed no surface metal–metal bonds are broken during adsorption. Eley gave the metal–hydrogen bond energy by an expression analogous to Eq. (2),

$$D_{MH} = (D_{MM} + D_{HH})/2 + 23.06(\chi_M - \chi_H)^2. \tag{6}$$

The problem is to find methods of evaluating D_{MM}, the metal–metal bond energy, and $(\chi_M - \chi_H)$, the difference in electronegativity that corrects for small polarization effects. Once these are accomplished, the bonding energy of a hydrogen atom to the metal surface, D_{MH}, may be calculated. By substituting Eq. (6) for D_{MH} into Eq. (5), we obtain a viable expression for calculating the heat of dissociative adsorption.

$$Q_0 = D_{MM} + 46.12(\chi_M - \chi_H)^2. \tag{7}$$

The metal–metal bond energy D_{MM} is estimated from the latent heat of sublimation λ. In a face-centered cubic metal, each atom has 12 nearest neighbors, and since each bond involves two atoms, we have the reasonable approximation

$$D_{MM} = \lambda/6. \tag{8}$$

Within the approximation of the calculation, the expression may be used for body-centered cubic lattices, where each atom has eight nearest and six next-nearest neighbors.

Three methods have been employed for determining electronegativities. The first and least satisfactory method utilizes Pauling's electronegativities estimated for diatomic molecules, which are determined in the following way. Pauling [3] defines a quantity Δ as the difference between the experimental bond energy and that expected for a pure covalent bond calculated from $(D_{AA} + D_{BB})/2$. The quantity Δ measures the resonance energy resulting from the ionic character of the unsymmetrical bond. From many experimental data, it was found possible to assign to the chemical elements electronegativity values (relative to unity for the hydrogen atom) such that their differences are approximately equal to the square roots of Δ in electron volts,

$$\Delta_{AB} = (\sqrt{\Delta_{AB}/23.06})^2 23.06 = (\chi_A - \chi_B)^2 23.06 \quad \text{kcal/mol.} \tag{9}$$

Values obtained in this way totally ignore the presence of a metal surface.

A second method, developed by Malone [4] for diatomic molecules, was adapted by Eley [2] to the chemisorptive bond. Malone found that the dipole moment of the bond in hydrogen halides, μ, expressed in electrostatic units (esu) was proportional to the difference in electronegativity given by Pauling, and that the proportionality factor was 1×10^{-18} esu. If μ is expressed in debyes ($1\text{ D} = 1 \times 10^{-18}$ esu), then the proportionality factor becomes unity and

$$\mu(\text{debyes}) = (\chi_H - \chi_X). \tag{10}$$

Eley applied this method to chemisorption on metals, calculating dipole moments from surface potentials available in the work of Mignolet [5–7]. A serious drawback exists. Values for surface potentials are available only at high coverages. But it is essential to relate electronegativities to dipole moments calculated from surface potentials measured at coverages approaching zero, where depolarization effects caused by dipole–dipole interactions are small. If depolarization effects are assumed to be small, the dipole moment at full monolayer coverage μ_f may be equated to the dipole moment at zero coverage μ_0 and the expression for full coverage used,

$$\mu_f = V/300(2\pi B) \tag{11}$$

where V is the surface potential in volts and B the number of sites per square centimeter. Otherwise, the equation given by Topping may be used,

$$\mu_0 = \mu_f[1 + (9\alpha/a^3)] \tag{12}$$

where α is the longitudinal polarizability of the bond and a is the lattice constant of the surface. For the adsorption of hydrogen atoms on metals, Eley writes

$$\mu_0(\text{debyes}) = (\chi_M - \chi_H). \tag{13}$$

A third method, developed by Stevenson [8], is based on Mulliken's

[9] electronegativity values for free atoms, which are defined by

$$\chi = (I + A_0)/2 \tag{14}$$

where I and A_0 are the ionization potential and electron affinity. Reasoning that electronegativity represents the attraction of a neutral atom in a stable molecule for electrons, Mulliken concluded that the average of the first ionization potential and the electron affinity should be a good measure of it. Pauling [3] showed that the sum of the ionization potential and the electron affinity for various metals was approximately proportional to his electronegativity values, χ_M, with a proportionality factor of 130 when expressed in kilocalories,

$$(I + A_0) = 130\chi_M \quad \text{kcal} \tag{15}$$

or

$$(I + A_0) = (130/23.06)\chi_M \quad \text{eV},$$

$$\chi_M = 0.177(I + A_0).$$

For bulk metals, both I and A_0 are set equal to ϕ, the work function of the metal, because the highest occupied and lowest unoccupied levels are at the Fermi surface. Stevenson writes

$$\chi_M = 0.177(2\phi) = 0.355\phi. \tag{16}$$

How do these simple models fit the experimental data? They must be judged on the basis of the heats of dissociative adsorption, Eq. (7), with $D_{MM} = \lambda/6$ from Eq. (8). We shall discuss only the Eley and the Stevenson equations, which become finally

$$Q_0 = \lambda/6 + 46.12(\mu_0)^2 \tag{17}$$

and

$$Q_0 = \lambda/6 + 46.12(0.355\phi - \chi_H), \tag{18}$$

respectively. Data are few and have been obtained chiefly using metal films, a difficult technique, although some data are available for metals supported on silica. For the Eley method, agreement between calculated and experimental heats of dissociative adsorption of hydrogen on various metals ranges from fair to poor, and there appears to be no trend in the discrepancies, which is evident in Table 2.1. The agreement using Stevenson's procedure appears somewhat better. It is moderately good for hydrogen on Fe, Co, Ni, Cu, Mo, Ta, and W; values for Pd and Pt are low; the value for Cr is very low, and the values for Ru, Rh, and Ir are too high. For the gases ethylene, oxygen, and nitrogen, agreement between observed and calculated values is poor without exception. We can conclude

TABLE 2.1

Comparison between Observed and Calculated Initial Heats of Chemisorption of Hydrogen and of Strengths of Hydrogen–Metal Bonds

Metal	Q_0 (kcal/mol) Obs.	Q_0 Calc. *a*	Q_0 Calc. *b*	D_{MH} (kcal/mol) Obs.	D_{MH} Calc.[b]
Cr	45	15.7	24	74.1	63.6
Fe	32–36	17	31.6	67.9–69.6	67.4
Co	24[c]	—	31.0	63.6[c]	67.1
Ni	29–32	17	28.9	66.1–67.6	66.0
Cu	28[c]	13.6	25.6	65.6[c]	64.4
Mo	40	—	42.9	71.6	73.0
Ru	26[c]	—	38.1	64.6[c]	70.6
Rh	26	23	32.3	64.6	67.7
Pd	27	—	22.5	65.1	62.9
Ta	45	32	49.6	74.1	76.4
W	45–52	44	45.6	74.1–77.6	74.4
Ir	26[c]	—	38.1	64.6[c]	70.6
Pt	28[c]	—	22.6	65.6[c]	62.9

[a] Eley [2].
[b] Stevenson [8].
[c] Supported metals; others, metal films.

only that neither the Eley method nor the Stevenson modification is useful for making faithful predictions of heats of chemisorption on metals.

With subsequent progress in research on chemisorption mechanisms, many reasons for the unreliability of these two approaches can be marshaled. First, however, we note that there is some theoretical justification for them, despite their empirical development. Koutecky [10] has shown that Eley's rule follows from the molecular orbital method for calculating the energy of a bonding electron in the chemisorption of an atom on the surface of a crystal, after making certain simplifying assumptions, including the assumption of complete localization of the electron. Mulliken had derived previously [11] by similar techniques the corresponding rule of Pauling for heterodiatomic molecules. We turn to the criticisms of Ehrlich [12, 13] who believes that the values of D_{MM} estimated from the heats of sublimation should be lower. He cites the work of Oriani [14] showing that the surface excess energy (energy to bring an atom from the interior to the surface) for liquid metals calculated from the heat of evaporation assuming pairwise bonding is high compared to values calculated from surface tension measurements and their temperature coefficients. Bonds in the surface layers were found to be about 13% stronger than those in the bulk, and this redistribution of binding energy for the surface implies a lower value of D_{MM} for use in calculating adsorption bonds. Similar considerations of crystalline filaments of copper, silver, and gold, close to their melting points, led to surface bonds about 25% stronger. Ehrlich [12] is especially concerned that the heat of chemisorption given by Eq. (7) is always positive, thus predicting that chemisorption is always exothermic. But endothermic chemisorption is possible when D_{MH} is less than $D_{HH}/2$, as is evident from Eq. (5). Using the geometric mean $(D_{MM}D_{HH})^{1/2}$ instead of the arithmetic mean $(D_{MM} + D_{HH})/2$, Ehrlich found agreement poorer for the transition metals, though chemisorption of hydrogen on Zn, Cd, and Hg were correctly predicted as endothermic.

There are more fundamental weaknesses that make revisions fruitless. Metal–metal bonds may be broken in some chemisorptions of hydrogen on transition metals, which would add an endothermic term. More than one hydrogen atom may be adsorbed per metal atom on the surface [15]. Conversely, an atom may occupy the central point of a lattice square of surface metal atoms with bonding to two or more of them [16]. Furthermore, chemisorbed hydrogen atoms are not always energetically equivalent. The existence of two or more states of adsorbed hydrogen, each having a different characteristic energy, has been demonstrated by many workers [17]. Heats of chemisorption are also known to fall precipitously with increasing coverage in many systems, indicating a pronounced and continuous surface energy distribution, which may depend strongly on how the surface is prepared. We must conclude that adaptations of Pauling's method to the determination of covalent surface bond strengths give an oversimplified picture of what is happening on an atomic scale.

Another empirical method which has been proposed correlates the heat of chemisorption and the heat of formation of the corresponding bulk compound. Inducement for such studies comes chiefly from the good linear correlation between heats of chemisorption of oxygen and heats of formation of the most closely corresponding bulk metal oxides [18–21]. But the metal–oxygen bond is believed to be ionic, and a better correlation might be expected for it than for the essentially covalent bond (see Chapter III). Tanaka and Tamaru [22] give a strictly empirical correlation by the expression

$$Q_0 = a[(-\Delta H_0°) + 37] + 20 \quad \text{kcal/mol} \tag{19}$$

where Q_0 is the heat of chemisorption at zero coverage, a is a constant depending on the gas, and $-\Delta H_0°$ is the heat of formation of the "highest" oxide per metal atom. A good linear correlation is obtained between the initial heat of chemisorption of oxygen and the heat of formation of the highest oxide. Inexplicably, the heat of formation of the highest *oxide* also correlates well with the heat of chemisorption of other adsorbates—H_2, N_2, NH_3, and C_2H_4—with different values of the constant a.

The notion that some single property of a metal determines its activity for chemisorption persists. Indeed, some justification exists in the empirical relationship, Eq. (19), and in the work of Trapnell [23], who finds that heats of chemisorption on different metals for the gases H_2, O_2, N_2, CO, CO_2, C_2H_4, and NH_3 follow a common pattern,

$$\text{Ti, Ta} > \text{Nb} > \text{W, Cr} > \text{Mo} > \text{Fe} > \text{Mn} > \text{Ni, Co} > \text{Rh} > \text{Pt} > \text{Pd} > \text{Cu, Au.} \tag{20}$$

The only definite reversal is the weak chemisorption of hydrogen on Mn. Eley's relationship suggests that there might be a correlation between the heats of chemisorption and either the heats of sublimation or the differences in electronegativities. The heat of sublimation is by far the larger factor and might be expected to dominate such a relationship. But the order of heats of sublimation is

$$\text{W} > \text{Nb, Ta} > \text{Mo} > \text{Rh} > \text{Pt} > \text{Ti} > \text{Co, Ni} > \text{Fe, Pd} > \text{Cr} > \text{Mn} > \text{Au} > \text{Cu,} \tag{21}$$

which bears no resemblance to the order of heats of chemisorption given earlier. The correlation between heats of chemisorption and electronegativities is no better. It is interesting to note that Gorbunov [24] derives theoretically a linear relation between the heat of chemisorption and the heat of sublimation as a first approximation.

2.2 Applications of the Valence-Bond Theory

The continued search for a single property of metals responsible for chemisorption led to the use of Pauling's valence-bond theory [3]. In this theory, cohesion in metals is assumed to originate from localized dsp hybrid bonds, differing from the electron-band theory, which pictures the metal as an array of positive nuclei through which valence electrons move almost freely. Since an atom in a metal lattice may have as many as 12 nearest neighbors (face-centered cubic), there may not be enough electrons to form localized bonds simultaneously between all nearest-neighbor pairs. To surmount this difficulty, Pauling assumes that electrons may resonate between pairs as explained below. Pauling studied Cr, Ni, and Co in detail, assuming in each case a covalence of six for the metallic state. The evidence, largely empirical, is briefly as follows. In going from left

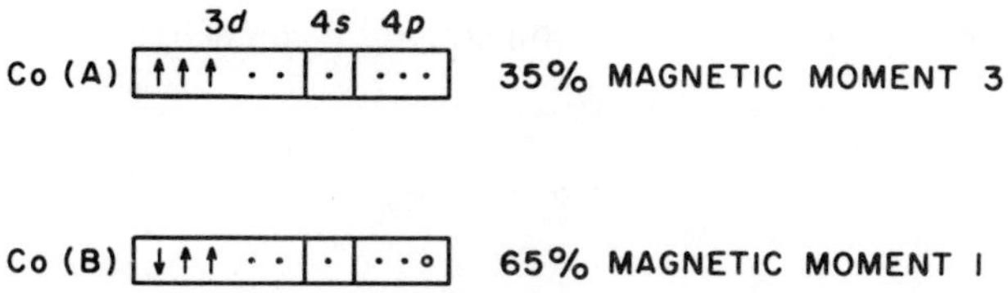

Fig. 2.1. Valence-bond representation of two forms of cobalt metal atoms.

to right across the sequences K–Cr, Rb–Mo, and Cs–W in the periodic table, the number of electrons available for bonding is found to increase and the atomic radii (thus the bond lengths) to decrease. With decrease in bond length and increase in number of bonding electrons per metal atom comes an increase in bond strength. But beyond Cr, Mo, and W, the atomic radii are roughly constant throughout the transition series. Pauling interprets this to mean that bond lengths, bond strengths, and number of bonding electrons are also constant. Since Cr has only six available electrons—five 3d and one 4s—he assumes that six is the common valence for all transition metals, though they may have more than six electrons in the d and s orbitals.

We use cobalt metal as an example and note that the problem is to put nine electrons into electronic orbitals, of which six are bonding electrons. Some of the electrons must be lifted to a higher state—the 4p state. Otherwise, the exclusion principle concerning the pairing of electrons would prohibit six unpaired electrons using only 3d and 4s orbitals. With five d, one s, and three p orbitals, nine are available altogether. Pauling sets up the following semiempirical rules governing the disposition of electrons and orbitals:

(a) ↑ or ↓ unpaired, nonbonding electrons, responsible for the magnetic moment, restricted to the d orbital;
(b) ↓ ↑ paired, nonbonding electrons, restricted to the d orbital;
(c) · bonding electrons, six available, filling orbitals in order of energy, d, s, p;
(d) ○ metallic orbitals, available locations for bonding electrons.

Using these rules, cobalt has six bonding electrons, three remaining electrons, of which 1.7 (magnetic moment in Bohr magnetons) are unpaired, giving 2.35 ↑ and 0.65 ↓ electrons for a net number of $2.35 - 0.65 = 1.7$ unpaired, nonbonding electrons. The number of occupied orbitals is $6 + 2.35 = 8.35$, leaving $9 - 8.35 = 0.65$ metallic orbitals. Fractional values may be considered as the average of two states, as shown in Fig. 2.1.

As mentioned earlier, Pauling assumes that the hybridized dsp bonds between pairs of metal atoms are continually resonating,

$$\begin{matrix} M{:}M \rightarrow M\ M \\ \quad\ \ \cdot\cdot\ \ \cdot\cdot \\ M{:}M \leftarrow M\ M \end{matrix}$$

a phenomenon he calls "synchronized" resonance. This picture may explain cohesion, but it does not explain, among other phenomena, the electrical conductivity of a metal. A switch in one pair of electrons is always balanced by a compensating switch in another pair, leaving no electrons for conduction. To account for electrical conductivity, a second type of resonance is postulated,

$$\begin{matrix} M{:}M & & M{:}M^- \\ & \leftrightarrows & \cdot\cdot \\ M{:}M & & M^+M \end{matrix}$$

called "pivotal" resonance. The metallic orbitals are used to accumulate extra electrons on atoms which receive an additional charge. A shift of one bond does not involve the simultaneous shift of another. Pivotal resonance qualitatively explains conduction.

Out of this picture of metals came a new quantity, the d character, which at one time [25] aroused great interest in connection with studies of chemisorption and catalysis. It is defined as the ratio of d bonding electrons to total bonding electrons plus metallic orbitals. For example, the d character of 35% Co(A) plus 65% Co(B) is $(2/6) \times 35 + (3/7) \times 65 = 39.4\%$. High d character means high participation of electrons in cohesion, and consequently could imply low availability for the formation of covalent chemisorptive bonds at the surface. If this is true, the order of heats of chemisorption should be the inverse of the order of d character, which is [23]

$$\text{Ti} > \text{Ta, Cr, Nb} > \text{Co} > \text{Fe} > \text{Ni, Mn} > \text{Mo} > \text{W} > \text{Pt} > \text{Pd} > \text{Rh} > \text{Cu, Au.}$$

The order does not agree well with the true order, Eq. (20).

We have mentioned factors previously which might obscure the role of d orbitals or d character in chemisorption, such as breaking of metal–metal bonds upon chemisorption, sharing of an adsorbate atom by more than one metal atom, more than one adsorbate atom per metal atom, and surface free valences differing from bulk orbitals. Nevertheless, the greater activity of d metals in general for chemisorption seems to imply that d orbitals must play at least a partial role. Recently, the problem has been compounded with the finding [16, 26, 27] that preformed hydrogen atoms can be chemisorbed on Group IB metals, whereas molecular hydrogen cannot. Perhaps chemisorption in some cases is limited by a high activation energy and not by the inability of the metal to chemisorb. There may not be as much difference as was once assumed between the potential energy curves of chemisorption of hydrogen atoms, for example, on sp (non-transition) and dsp (transition) metals. As Dowden [28] suggests, there may be an initial state of chemisorption intermediate between physical adsorption and strong chemisorption, and there may be an important difference between the potential energy curves for this weak chemisorption. For transition metals, this potential curve may intersect the curve for strong chemisorption in such a way that very little activation energy is required, in contrast to the large activation energy requirement for non-transition metals.

We conclude this section with a few speculations on the detailed nature of the covalent chemisorptive bond in the light of the valence-bond theory. Invoking this theory, Dowden [28] reasons that there are at least two types of chemisorption on metals: (1) a weak form involving d orbitals and (2) a strong form using dsp bonds. The latter implies a resemblance between chemisorptive bonds and the cohesive bonds in bulk metal, lending qualitative support to the Eley–Stevenson approach (see Section 2.1). In weak chemisorption, Dowden believes that atoms or molecules during chemisorption will overlap first with d_{z^2}, then with (d_{zx}, d_{zy}), and finally with $d_{x^2-y^2}$ and d_{xy}, both of which lie in the surface. Bond strengths along the z axis will be controlled by the fraction of each orbital available, its energy, and its electron content. The d_{z^2}, d_{zx}, and d_{zy} orbitals project from the surface, but can form only weak π bonds. Strong chemisorption bonds are considered to be similar to the dsp-bonding hybrids of the bulk metal. It may be that chemisorption bonds are drastically altered as coverage increases. At low coverages, dsp bonds at the surface of a d metal must be essentially free, otherwise the Eley–Stevenson equation for heats of chemisorption discussed in Section 2.1 should contain an endothermal term for breaking of metal–metal bonds. Dowden further postulates that these free dsp bonds provide a supply of metallic orbitals, little used in pivotal resonance of metal–metal bonds, and available for chemisorption. These bonds will be used up as coverage increases. Heats of chemisorption will then fall, because there will be increasing interference with pivotal resonance in the metal until the point is reached when an endothermal term for breaking metal–metal bonds must be included in any calculation of the heat of adsorption. The fact [14] that pairwise bonding of surface metal atoms is stronger than in the bulk implies a difference between surface and bulk dsp hybrid orbitals. Dowden [28] postulates that nearest-neighbor bonds in α iron could rearrange to give surface states as follows:

$$d^3sp^2 \rightarrow d^2p \text{ (trigonal pyramid)} + dsp \text{ (bent planar)};$$

and for next-nearest neighbors:

$$d^2sp^2 \longrightarrow d^2sp^2 \text{ (distorted tetragonal pyramid)}.$$

On semiconducting oxides, weak chemisorption may proceed as on metal surfaces, using metal d orbitals. Strong chemisorption is probably sufficiently ionic, though possessing some covalent character, for us to apply the boundary-layer theory (Chapter 5). Insulators operate on the same principle but with a stronger affinity for electronegative species. Brönsted-acid centers gradually give way to Lewis-acid centers at elevated temperatures with loss of water. In place of d or dsp orbitals, the residual valences are now either p_z orbitals, providing centers for weak chemisorption, or sp hybridized orbitals for strong chemisorption. These centers may accept lone pairs or π electrons in Dewar bonds, but they cannot react with hydrogenlike molecules as easily as d and dsp orbitals.

2.3 Formal Quantum-Mechanical Calculations

Most of the quantum-mechanical studies relevant to this chapter deal with chemisorption on metals. Therefore, a few preliminary remarks about the theory of metals are appropriate. We shall give a more extensive treatment in Chapter 8, which is not limited to localized covalent bonding.

A complete description of gas–metal interactions requires a detailed understanding of both the metal and the gas particle. In the foreseeable future, we must get along with severe approximations for the metal. Two basic approaches [29] are available: (1) the nearly-free-electron model; and (2) the localized model, in which electrons are essentially localized either on atoms or in bonds. The models represent extremes, and each has its area of utility. The nearly-free-electron model explains most of the bulk effects and properties of metals in terms of a Fermi sea of electrons in the periodic field of ion cores, taking into account the interactions of the electrons with themselves and with the ion cores. Even the classical free-electron theory, developed by Lorentz, Drude, and Debye around the turn of the century, and which regards the valence electrons in a metal as noninteracting particles of an ideal gas, brings out the general features of such phenomena as electrical and thermal conductivity. Introducing the quantum-mechanical treatment of electrons in a box, the exclusion principle, and Fermi–Dirac statistics allows meaningful calculations of other quantities, such as electronic specific heat, mean free path, and diamagnetic and paramagnetic susceptibilities. When structure is given to the metal—the periodic field of ion cores—the results of calculations are even more satisfying and useful. In models of this type, the surface is represented by a potential barrier and the potential variation in its plane is ignored—a serious omission, especially when dealing with surface phenomena.

The other extreme, the localized model, fits the phenomenon of chemisorption better. The surface is approximated by a single atom or an array of a few atoms. These atoms may be assigned the electronic structure of isolated metal atoms or atoms in the metallic state whose orbitals are often fractionally occupied. With this model for the metal, we think of chemisorption as the formation of a localized surface complex perturbed by the rest of the crystal. Experimental studies may also be divided into corresponding approaches: the study of bulk properties or the direct study of surface interactions. We may measure the bulk properties of metals, such as electrical conductivity and magnetic properties, and indirectly draw conclusions concerning surface properties; or we may examine surface processes directly. The latter studies are now being pursued vigorously, and are leading to information on the direct interactions of various particles with metallic surfaces. Among the methods employed for surface examinations are infrared spectroscopy; Auger spectroscopy; ion-neutralization spectroscopy; energy-distribution measurements of field-emitted electrons and of inelastically scattered slow electrons; low-energy electron diffraction (LEED); field-ion microscopy; field-electron microscopy; field desorption experiments; molecular-beam scattering; photochemistry of surface layers; photoemission; slow electron and ion desorption experiments; and adsorption calorimetry. In addition, certain well-directed classical investigations of chemisorption point to the role of individual surface atoms, notably the work of Sachtler and van der Plank [30] on the chemisorption of hydrogen by nickel–copper alloys.

All modern theoretical approaches to localized chemisorption are based on molecular orbital theory, a technique for combining the atomic orbitals of a group of atoms into molecular orbitals. They are usually combined in linear fashion, and the general method is known as the linear combination of atomic orbitals (LCAO–MO). The wave function for the molecular orbital is

$$\psi = \sum C_r \phi_r, \tag{22}$$

which includes a small, finite number of atomic orbitals of metal surface atoms and atomic orbitals of the adsorbate molecule. The wave equation

$$H\psi = E\psi, \tag{23}$$

where H is the one-electron Hamiltonian operator and E is the orbital energy, is multiplied by ψ and put into the form

$$E = \int \psi H \psi \, d\tau \Big/ \int \psi^2 \, d\tau. \tag{24}$$

After substituting ψ from Eq. (22) into the energy equation, Eq. (24), the energy is obtained by the variation method, in which the energy is minimized with respect to each coefficient C_r:

$$E = \int \sum_r C_r \phi_r H \sum_s C_s \phi_s \, d\tau \Big/ \int \sum_r C_r \phi_r \sum_s C_s \phi_s \, d\tau \tag{25}$$

$$(\partial E / \partial C_r) = 0 \tag{26}$$

A set of secular equations is obtained of the form

$$\sum_s (H_{rs} - E S_{rs}) C_s = 0, \qquad r = 1, 2, 3, \ldots, \tag{27}$$

which is solved for the C_s and E by the determinant

$$| H_{rs} - E S_{rs} | = 0 \tag{28}$$

where

$$H_{rs} = \int \phi_r H \phi_s \, d\tau \tag{29}$$

and S_{rs} is the overlap integral

$$S_{rs} = \int \phi_r \phi_s \, d\tau. \tag{30}$$

There are many techniques for evaluating the H_{rs} and S_{rs} or simplifying the energy equation so that it will yield numerical answers. A common procedure is the "tight binding" approximation, in which S_{rs} is set equal to zero when $r \neq s$ and to unity when $r = s$ [31]. Another method which has become popular since its development in 1963 by Hoffmann [32] is called the extended Hückel molecular orbital method (EHMO). It retains all the H_{rs} and S_{rs}, using semiempirical methods for evaluating them.

We note in passing that there is nothing in the LCAO–MO method which limits it in principle to the use of a finite sum of orbitals. In fact, we shall make use of infinite summations of atomic orbitals for the interaction of an adsorbed atom with semi-infinite one-, two-, and three-dimensional crystals in later chapters. Using such summations leads to band structures, the possibilities of free surface states, and localized or nonlocalized chemisorption states.

In the remainder of this section, we give a few examples of calculations which have been reported. Anderson and Avery [33] have calculated electronic energies by the simple Hückel MO treatment for 1,3 diadsorbed butane on platinum, which they assume to be the precursor of isomerization. They assume a pair of platinum orbitals suitable for π bonding to the p_z atomic orbital of a carbon atom. Over a range of assumed values for the overlap integral S, they found the bonding electrons were uniformly distributed between the platinum and carbon atoms. In other words, the chemisorptive bond is covalent.

Other workers [34, 35], employing idealized systems, have investigated small finite models, comparing them with semi-infinite models. For example, Blyholder and Coulson [34] showed that small finite models of one- and two-dimensional crystals predict the same surface states as semi-infinite models. They used both the simple Hückel MO treatment and a more sophisticated LCAO–MO approach on a six-atom linear chain where each atom has one valence electron of s type. Both models gave a reasonably good description of the orbital changes that occur as the electron-attracting power of an end atom is changed, thus simulating chemisorption.

An interesting paper by Robertson and Wilmsen [36] uses the EHMO model and reports calculations of adsorption energies for several organic molecules at various locations on a Pb surface. In making these calculations, they made the following assumptions for the model 100 plane of the Pb surface and the entering group of adsorbate atoms:

1. The important interactions will be between the surface atoms and the entering group, the contributions of the underlying atoms being an order of magnitude less for all interactions.
2. The five d orbitals of the lead atom are not involved in bonding to the entering group, since the chemical activity of Pb is due to the s and p orbitals, and d orbitals are not involved in the more common reactions.
3. Inasmuch as orbital overlap is very small at distances greater than 4 Å, the entering group will be unaffected by surface atoms beyond the nearest neighbors of the central atom.

Two types of adsorbate groups were used: (1) electron-rich, such as methyl mercaptan and methyl amine; and (2) electron-poor, such as the acetoxy radical, the ethyl radical, and the Pb atom. With the electron-rich group, the system energy was always greater than the energy at infinite separation, indicating repulsion and no chemisorption. Conversely, the system energy for the electron-poor group remained below the energy at infinite separation at all distances greater than about 2 Å, indicating that chemisorption occurs. The calculations are in agreement with the known chemical facts that lead does not react with nonoxidative reagents (electron-rich), but does react with alkyl radicals and other lead atoms (electron-poor). Estimating an equilibrium distance of about 2.5 Å, they calculated the binding energy for the ethyl radical to be 0.6 eV, for the acetoxy radical 4 eV, and for the Pb atom 1.58 eV. The latter figure compares favorably with the atomization energy for Pb, 1.99 eV, and lends credibility to the EHMO calculations.

Finally, we cite a series of papers by Roev *et al.* [37–40], who calculated energies for nitrogen, hydrogen, and nitric oxide complexes on a number of metals; and the work of Bonch-Bruevich [41, 42] and Kogan and Sandomirsky [43], who calculated the localized energies of adsorption of hydrogenlike atoms on crystal defects.

2.4 Covalent Surface Complexes

Out of the multitude of experimental papers on the nature of the chemisorptive bond, we shall discuss only a few. Techniques range from classical adsorption experiments to the use of a growing number of sophisticated modern instruments. Evidence on the nature of a bond varies from meager to extensive. No system has been explained unequivocally yet, not even qualitatively. Attacking a chemisorptive system with several techniques, as a few laboratories are now doing, will help. Also, better coordination of theoretical and experimental approaches would help.

Simple Gases on Metals and Metal Oxides

The chemisorption of hydrogen on tungsten is far more complex than once believed. Tamm and Schmidt [44], in flash desorption studies, found two atomic states of hydrogen on the 110 face of tungsten and four states on the 111 face. At saturation, the ratio of hydrogen to tungsten atoms on the 110 face is unity and on the 111 face it is 4. The tightly bound states have the same energy on all faces, but the weakly bound states on the 111 face with energies of 14 and 22 kcal/mole are absent on other faces. It appears that the bonds are highly specific to the crystal plane. Multiple states on a face imply highly directional bonds; otherwise, only one state would exist, as predicted by the hard-sphere model. Further conclusions are made in terms of crystal-field theory, which explains the splitting of the tungsten d-band energy levels and their highly directional character (see Chapter 10).

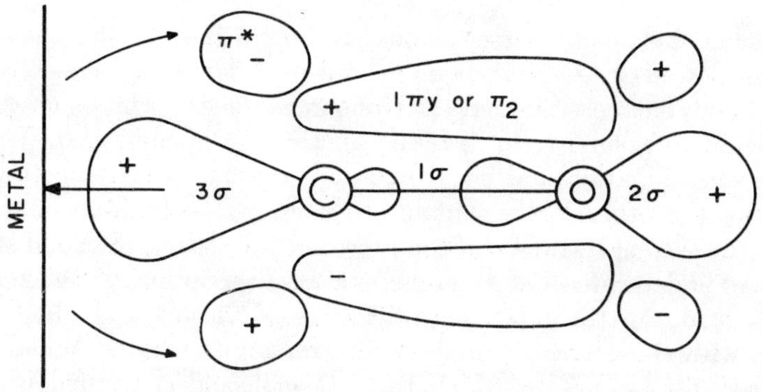

Fig. 2.2. The molecular orbital description of gaseous carbon monoxide. Arrows indicate the possible electronic interactions that may occur on a transition metal [52].

Infrared studies by Kokes and co-workers [45–50] and Eischens *et al.* [51] reveal that there is a fast chemisorption of hydrogen on zinc oxide during which new bands appear at 3510 and 1710 cm^{-1}. These are associated with the formation of OH and Zn–H. A slow adsorption of hydrogen occurs without the appearance of any new bands in the infrared spectra. Eischens *et al.* [51] believe this adsorption corresponds to the formation of H^+.

Carbon monoxide is an exceptionally good molecule for chemisorption studies because of its high stability. With a bond strength of 256.7 kcal/mole, it offers one of the few examples of completely nondissociative chemisorption. We give a general molecular orbital description of the CO molecule and its bonding to a transition metal atom, before considering specific metal–CO systems. The 2s and the $2p_x$ orbitals on both the carbon and oxygen atoms form sp hybrids (Fig. 2.2). One pair overlaps along the line of centers of the carbon and oxygen atoms forming the σ bond of the molecule, labeled 1σ in the figure. The remaining two orbitals (2σ and 3σ) are oriented away from each other, one on the carbon and one on the oxygen atom, and accommodate the nonbonded lone pairs of electrons on each atom. Two degenerate π bonds formed by the overlap of $2p_y$ and $2p_z$ orbitals of carbon and oxygen complete the molecular orbital picture of CO. The 3σ orbital on carbon extends farther than the 2σ orbital on oxygen from its atom center, but is offset by the asymmetry of the π orbitals, in which an electron has a greater probability of being near the oxygen than near the carbon nucleus by a factor of 4. In this way, the very small dipole moment of CO is explained.

The first step in the bonding of the CO molecule to a transition metal atom may proceed by donation of the lone pair on the carbon atom into vacant d orbitals of the metal atom—a more probable step than donation by oxygen, whose lone pair electrons are bound tighter. But the electron-donating power of CO is also small, and in order to stabilize the metal–carbon bond, back donation of electrons from filled d orbitals on the metal into vacant antibonding π^* orbitals on the CO is envisaged. These two mechanisms, donation and back donation, exert a synergistic effect, each enhancing the other in its attempt to counter the buildup of electronic charge, thus strengthening the bond beyond the sum of the bond strengths calculated independently. A more detailed description of the metal–CO bond depends on the relative availability of vacant and filled metal d orbitals with the correct symmetry and range of energies. When we consider that the species to which the CO molecule is bonded is a metal surface instead of an isolated atom, the whole question of metal band structure and localized versus nonlocalized bonding is raised. But it makes little difference at this point how the problem is viewed; the results of this section are not sufficiently detailed to warrant the distinction. Furthermore, the symmetry and directional characteristics of d-band levels produced by crystal-field splitting (see Chapter 10) and d orbitals of isolated metal atoms may be considered qualitatively the same [53].

Carbon monoxide is chemisorbed on tungsten in two different states, α and β, so designated by Hickmott and Ehrlich [54] on the basis of flash filament desorption studies. The α state desorbs between 200 and 500°K, and the strongly held β state desorbs between 1100 and 1850°K. The formation of α-CO becomes significant only when the uptake of β-CO starts to decrease. Later studies [55, 56] resolved the β-CO flash desorption peak into three distinct substates, β_1, β_2, and β_3. Since the desorption kinetics are first order, it was concluded that chemisorption was nondissociative in all three substates. Redhead [55] suggests that α-CO chemisorption occurs on single sites statistically isolated by the random, immobile, two-site adsorption of β-CO. Monte Carlo statistics indicate that about 13% of the total sites are isolated in this way, a figure which agrees well with Redhead's findings. There is general agreement that the α-CO adsorption complex is attached to the metal through the carbon atom (Fig. 2.3c), although more than one bonding energy exists in this category. The β-CO chemisorption occurs by a two-site mechanism according to Ehrlich and Redhead, both of whom suggest that the multiplicity of this state may be accounted for by adsorption on different crystallographic planes with different energies and lattice spacings. Redhead [55] makes a further suggestion that the structure of β-CO is a bridged carbonyl (Fig. 2.3a) in which CO orbitals overlap the d orbitals of two metal atoms, as for example in the carbonyl molecule $Fe_2(CO)_9$. Controversy still exists over the bridged structure. In addition to flash filament desorption

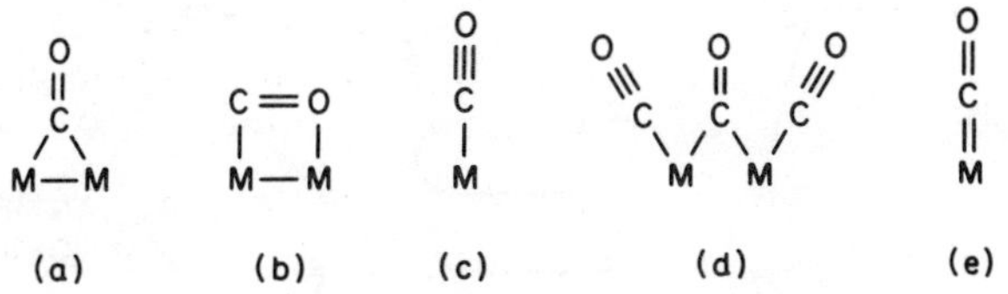

Fig. 2.3. Suggested adsorption complexes for CO on tungsten.

studies, other techniques are being applied: field emission studies, field-ion microscopy, LEED, and electron impact desorption studies. The last technique has shown the absence of CO^+ and C^+ [57].

By flash filament [58, 59] and LEED studies [60], the close similarity between adsorption of CO on tungsten and molybdenum has been shown. Very little information is available for the CO–chromium system.

For CO adsorption on Group VIII metals supported on materials such as silica and alumina, Eischens and Pliskin [61] in a review of infrared studies report two CO–metal structures: a bridged carbonyl-type structure (Fig. 2.3a) and a linear monodentate structure (Fig. 2.3c). Blyholder [62, 63] takes issue with the bridged structure, preferring a linear structure chemisorbed on edge or corner sites. Garland *et al.* [64] have observed the infrared spectra of CO adsorbed on metal films instead of supported metals of Group VIII. On Ni, Pt, Pd, and Rh films their results are qualitatively similar to those on supported samples. In one case, however, Yang and Garland [65] interpreted infrared spectra of CO adsorbed on alumina-supported rhodium as representing two linear structures attached to a single rhodium site, a structure not observed on rhodium films [64]. A careful analysis of infrared spectra of CO adsorbed on silica-supported platinum leads Lin *et al.* [66] to propose a linear structure with a double bond between the platinum and carbon (Fig. 2.3e).

Olefin Complexes

The initial heat of chemisorption of ethylene on zinc oxide was found to be 14 kcal/mole [50] and the infrared bands correspond to the fundamentals for gaseous ethylene plus several bands forbidden in the gaseous state. Unlike propylene, there is no dissociative adsorption. The facts that the infrared spectra exhibit olefinic character and that the heat of adsorption is appreciable suggest π bonding. For steric reasons, it is presumed that ethylene is bonded to oxygen-ion sites which obscure the much smaller zinc ions [50]. Nuclear magnetic resonance studies confirm the olefinic structure of adsorbed ethylene [67].

On catalysts for the disproportionation of olefins [68, 69], such as

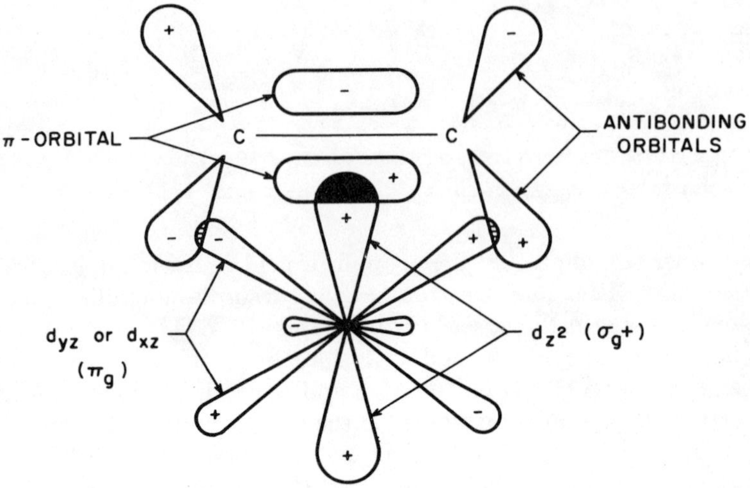

Fig. 2.4. Pi-adsorbed olefin on a transition metal.

oxides of molybdenum, tungsten, and rhenium supported on silica or alumina, olefins are again presumed to form π-bonded complexes, as illustrated in Figure 2.4, but in this case on the metal ion. The reaction is believed to proceed in the adsorbed state through a quasi-cyclobutane intermediate [69, 70] with metal d orbitals overlapping orbitals of reactant, intermediate, and product,

$$\begin{matrix} R_1CH_2CH{=}CHR_2 \\ R_1CH_2CH{=}CHR_2 \end{matrix} \leftrightarrows \begin{matrix} R_1CH_2CH{-}{-}{-}CHR_2 \\ \vdots \qquad\quad \vdots \\ R_1CH_2CH{-}{-}{-}CHR_2 \end{matrix} \leftrightarrows \begin{matrix} R_1CH_2CH \\ \| \\ R_1CH_2CH \end{matrix} + \begin{matrix} HCR_2 \\ \| \\ HCR_2 \end{matrix} \tag{31}$$

where R_1 and R_2 are alkyl groups or hydrogen atoms. Kinetic studies show that the reaction is first order with respect to catalyst, which implies that two olefin molecules are bound to one metal site. Lewandos and Pettit [71] point out that the course of the disproportionation reaction in the gas phase through a cyclic intermediate is forbidden by the Woodward–Hoffman symmetry rules [72]. But the process is allowed in the adsorbed phase with overlap of metal d orbitals.

We come now to one of the most important processes, both industrially and scientifically—the Ziegler–Natta process for the polymerization of olefins. Several schemes have been put forth for the mechanism of the reactions [73–81] and, with one exception [75], they agree that polymer growth occurs by insertion of a monomer molecule between a metal atom and a carbon atom. The most detailed theory is that of Cossee and Arlman [82–85], who describe the chemisorbed state of the reacting olefin, the intermediates, and the product. A common type of catalyst is the reaction product between $TiCl_3$ and aluminum triethyl. Arlman [83] believes that the availability of metal ions with vacant positions of ligand coordination is a critical factor for polymer chain growth, and presents arguments indicating that the surfaces of transition metal chlorides (e.g., $TiCl_3$) carry such vacancies. The formation of an active center on the surface

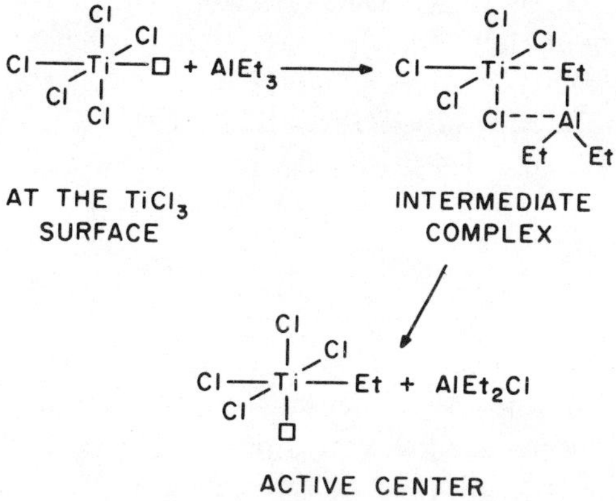

(32)

is pictured as in Eq. (32). Cossee [82] adopts the ideas of Chatt and Duncanson [86], which involve the formation of a π complex as shown in Fig. 2.4. As mentioned earlier, the propagation step is believed to take place by insertion of an olefin molecule between a metal atom and an alkyl group. Thus, the transition metal ion must have an alkyl ligand and the possibility to coordinate an olefin molecule as depicted in Eq. (32). When an olefin like ethylene is brought into such a configuration, the metal d_{yz} orbital (Fig. 2.4), originally having an energy equal to that of the d_{xy} and d_{xz} orbitals, combines with the π-antibonding orbital of ethylene, forming an orbital of considerably lower energy. The energy gap between this orbital and the orbital representing the bond between the alkyl group and the transition metal is significantly reduced. Since the promotion of an electron from the metal–alkyl bonding orbital by thermal excitation is now easier, the alkyl–metal bond will be more susceptible to breaking

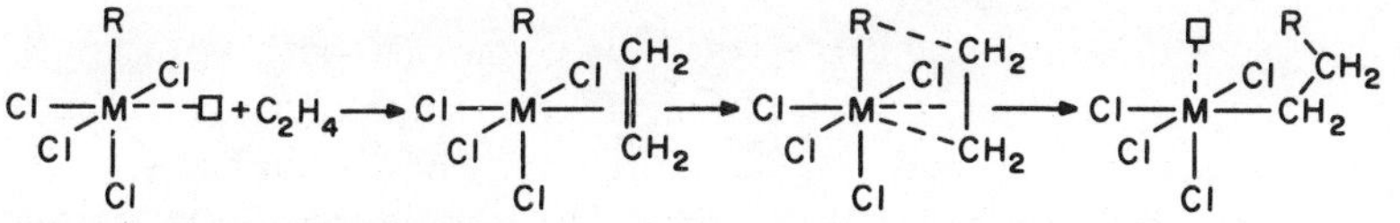

(33)

into radicals. The steps are shown in Eq. 33. Using propylene and higher olefins, the process forms stereospecific polymers, whose formation the theory also explains [84].

Not all surfaces chemisorb olefins as π complexes. Friedel–Kraft catalysts, for example, are believed to chemisorb olefins end-on through an electron pair removed from the π-molecular orbital of the olefin molecule [87]. A similar situation probably exists in the chemisorption of olefins on silica–alumina [88]. When 1-butene is added to silica–alumina, a strongly adsorbed complex is formed which varies from monomer to tetramer with increasing pressure. The initial heat of adsorption of essentially 1-butene monomer is approximately 8 kcal/mole. If this low heat is a measure of the adsorption bond strength, one would not expect the polymer subsequently formed on a site to be irreversibly adsorbed. In order to explain the results, the mode of adsorption may be pictured as follows [87]:

$$\begin{array}{lr} \mathrm{C::C} \rightarrow \mathrm{:\bar{C}:C^+} & +40\ \mathrm{kcal} \\ \mathrm{S^+ + :\bar{C}:C^+} \rightarrow \mathrm{S:C:C^+} & -48\ \mathrm{kcal} \\ \hline \mathrm{S^+ + C::C} \rightarrow \mathrm{S:C:C^+} & -8\ \mathrm{kcal} \end{array} \tag{34}$$

where S^+ represents an adsorption site and C::C represents an olefin. The endothermic heat of the first reaction (+40 kcal/mole) is a quantum-mechanical estimate made by Harman and Eyring [89]. The heat of the second reaction (−48 kcal/mole) is calculated using the overall observed heat of adsorption given by the third equation, −8 kcal/mole. The strength of the chemisorption bond based on these assumptions is given by the second equation (−48 kcal/mole), which is sufficiently large to account for the observed irreversible adsorption. A mechanism of this type could explain net endothermic heats of adsorption. If adsorption occurs directly through the π bond, irreversible adsorption would more likely be accompanied by a large observed heat of adsorption.

Other Complexes

The stepwise dissociative chemisorption of ethane on many transition metals was once thought to proceed through the conversion of adsorbed ethyl radicals to adsorbed ethylene and perhaps even further. By using deuterium tracer techniques, the results were nicely explained [90] by the α,β diadsorption of ethylene

$$\begin{array}{cc} \mathrm{H_2C} & \!\!\!\!-\!\!\!\!\!\! & \mathrm{CH_2} \\ | & & | \\ \mathrm{M} & & \mathrm{M} \end{array} \tag{35}$$

But this classical α,β mechanism, as it was called, soon ran into difficulties. For instance, the exchange of propane, n-pentane, and n-hexane with deuterium on palladium films [91] revealed *initial* deuterium distributions, produced in one sojourn on the surface, where the perdeutero isomer accounted for more than 80% of the total products, while under the same conditions ethane gave only about 50%. To explain the initial formation of products of high deuterium content, an extended α,β mechanism would have to be invoked in which the possibility of triadsorbed, tetraadsorbed, and other multiadsorbed species are considered. However, such tri- and tetra- and other multiadsorbed species are highly improbable when only σ-bonded adsorbates are considered. The geometric restrictions imposed on a paraffinic molecule σ-bonded through sp^3 hybrid, tetrahedrally oriented, carbon atom orbitals are too severe. Another difficulty encountered in the classical α,β mechanism is the explanation of the initial deuterium exchange of all ten hydrogen atoms in cyclopentane on palladium. By the classical α,β mechanism, only five hydrogen atoms should exchange initially; then the molecule must "turn over" and readsorb in order to exchange five more hydrogen atoms on the other side of the planar ring.

To circumvent the difficulties, the classical α,β mechanism was dropped, and a multicentered bonding to the same metal atom was considered in its place. Intermediates were proposed that underwent stepwise interconversion as σ- and π-bonded ligands to one atomic center on a metal surface [92]. These are illustrated in (36) for the stepwise interconversion of σ- and π-bonded ligands of chemisorbed cyclohexane and its unsaturated derivatives.

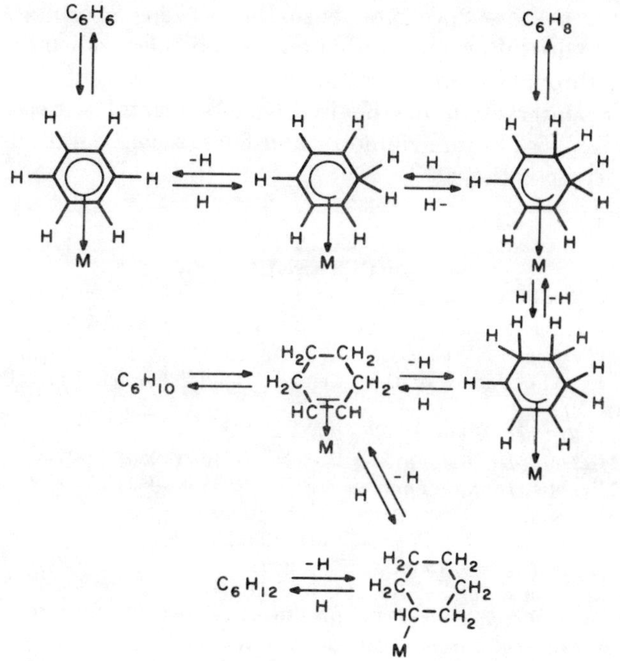

(36)

Finally, we describe chemisorption on the Brönsted and Lewis sites on

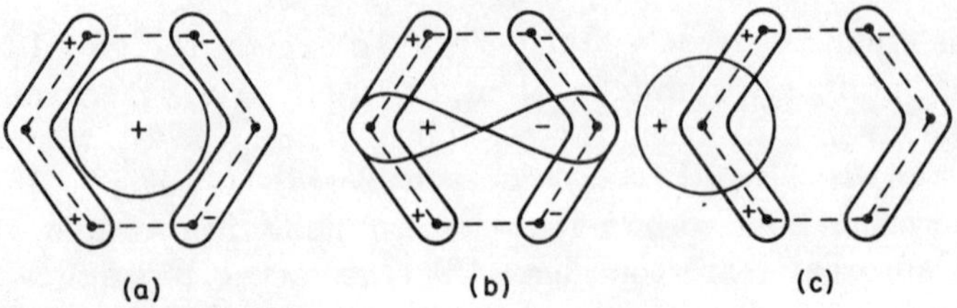

Fig. 2.5. Models of the benzene–hydroxyl bond on silica–alumina [99].

adsorbents such as silica–alumina. Basic substances like pyridine and ammonia with unshared pairs of electrons are bonded according to infrared evidence in two ways [61, 93–98]: coordinately on Lewis sites, and as pyridinium or ammonium on Brönsted sites. On typical commercial silica–alumina catalysts, Basila and Kanter [96] determined that the ratio of Lewis to Brönsted sites varied in the range 3 to 6. On alumina, only Lewis sites exist. Cusumano and Low [99] speculated on the nature of the benzene–hydroxyl (Brönsted) bond on silica–alumina. They suggest three models: (1) the hydroxyl proton located on the axis of symmetry perpendicular to the plane of the benzene ring, interaction of the hydrogen s orbital with the benzene π orbital, Fig. 2.5a; (2) symmetrical interaction of the hydrogen atom 2p orbital with the benzene π orbital, Fig. 2.5b; and (3) the hydroxyl proton located on the axis perpendicular to the ring and passing through a carbon atom or through the center of a carbon–carbon bond, interaction of the hydrogen s orbital with the benzene π orbital, Fig. 2.5c. From symmetry and energy considerations, the most favored structure is the last.

REFERENCES

1. L. Pauling, *J. Amer. Chem. Soc.* **54,** 3570 (1932).
2. D. D. Eley, *Discuss. Faraday Soc.* No. 8, 34 (1950).
3. L. Pauling, "The Nature of the Chemical Bond." Cornell Univ. Press, Ithaca, New York, 1939.
4. M. Malone, *J. Chem. Phys.* **1,** 197 (1933).
5. J. C. P. Mignolet, *Discuss. Faraday Soc.* No. 8, 105 (1950).
6. J. C. P. Mignolet, *Discuss. Faraday Soc.* No. 8, 326 (1950).
7. J. C. P. Mignolet, *Rec. Trav. Chim.* **74,** 701 (1955).
8. D. P. Stevenson, *J. Chem. Phys.* **23,** 203 (1955).
9. R. S. Mulliken, *J. Chem. Phys.* **3,** 573 (1935).
10. J. Koutecky, *Trans. Faraday Soc.* **54,** 1038 (1958).
11. R. S. Mulliken, *J. Chim. Phys.* **46,** 497 (1949).
12. G. Ehrlich, *Proc. Int. Congr. Catal., 3rd* **1,** 113 (1964).
13. G. Ehrlich, *J. Chem. Phys.* **31,** 1111 (1959).

14. R. I. Oriani, *J. Chem. Phys.* **18,** 575 (1950).
15. S. F. Adler and J. J. Kearney, *J. Phys. Chem.* **64,** 298 (1960).
16. R. V. Culver, J. Pritchard, and F. C. Tompkins, *Z. Elektro-chem.* **63,** 741 (1959).
17. G. C. Bond, "Catalysis by Metals," Chapter 5. Academic Press, New York, 1962.
18. N. W. Roberts, *Nature* (*London*) **188,** 1020 (1960).
19. D. Brennan, D. O. Hayward, and B. M. W. Trapnell, *Proc. Roy. Soc.* (*London*) **A256,** 81 (1960).
20. W. M. H. Sachtler and L. L. van Reijen, *Shokubai* (*Tokyo*) **4,** 147 (1962).
21. G. C. Bond, "Catalysis by Metals," p. 472. Academic Press, New York, 1962.
22. K. Tanaka and K. Tamaru, *J. Catal.* **2,** 366 (1963).
23. B. M. W. Trapnell and D. O. Hayward, "Chemisorption," p. 210. Butterworth, London and Washington D.C., 1964.
24. A. I. Gorbunov, *Dokl. Akad. Nauk SSR* **161,** 1121 (1964).
25. O. Beeck, *Discuss. Faraday Soc.* No. 8, 118 (1950).
26. J. Pritchard and F. C. Tompkins, *Trans. Faraday Soc.* **56,** 540 (1960).
27. R. V. Culver, J. Pritchard, and F. C. Tompkins, *Proc. Int. Conf. Surface Activity* **2.** Butterworth, London and Washington, D.C., 1957.
28. D. A. Dowden, "Chemisorption" (W. E. Garner, ed.), p. 9. Butterworth, London and Washington D.C., 1958.
29. Z. Knor, *J. Vac. Sci. Technol.* **8,** 57 (1971).
30. W. M. H. Sachtler and P. van der Plank, *Surface Sci.* **18,** 62 (1969).
31. T. B. Grimley, *Advan. Catal.* **12,** 1 (1960).
32. R. Hoffmann, *J. Chem. Phys.* **39,** 1397 (1963).
33. J. R. Anderson and N. R. Avery, *J. Catal.* **7,** 315 (1967).
34. G. Blyholder and C. A. Coulson, *Trans. Faraday Soc.* **63,** 1782 (1967).
35. H. H. Dunken and Ch. Opitz, *Z. Phys. Chem. Neue Folge* **60,** 25 (1968).
36. J. C. Robertson and C. W. Wilmsen, *J. Vac. Sci. Technol.* **8,** 53 (1971).
37. L. M. Roev and V. A. Zasukla, *Dopov. Akad. Nauk. Ukr. RSR Ser. B* **32,** 434 (1970).
38. L. M. Roev and V. A. Zasukla, *Teor. Eksp. Khim.* **6,** 614 (1970).
39. V. A. Zasukla and L. M. Roev, *Teor. Eksp. Khim.* **7,** 8 (1971).
40. L. M. Roev and V. A. Zasukla, *Zh. Fiz. Khim* **45,** 322 (1971).
41. V. L. Bonch-Bruevich, *Zh. Fiz. Khim.* **27,** 662 (1953).
42. V. L. Bonch-Bruevich, *Zh. Fiz. Khim.* **27,** 960 (1953).
43. S. M. Kogan and V. B. Sandomirsky, *Bull. Acad. Sci. USSR Ser. Chem.* 1681 (1959).
44. P. W. Tamm and L. D. Schmidt, *J. Chem. Phys.* **54,** 4775 (1971).
45. W. C. Connor, R. A. Innes, and R. J. Kokes, *J. Amer. Chem. Soc.* **90,** 6865 (1968).
46. W. C. Connor and R. J. Kokes, *J. Phys. Chem.* **73,** 2436 (1969).
47. A. L. Dent and R. J. Kokes, *J. Amer. Chem. Soc.* **91,** 7207 (1969).
48. A. L. Dent and R. J. Kokes, *J. Phys. Chem.* **73,** 3772 (1969).
49. A. L. Dent and R. J. Kokes, *J. Phys. Chem.* **73,** 3781 (1969).
50. A. L. Dent and R. J. Kokes, *J. Phys. Chem.* **74,** 3653 (1970).
51. R. P. Eischens, W. A. Pliskin, and M. J. D. Low, *J. Catal.* **1,** 180 (1962).
52. R. R. Ford, *Advan. Catal.* **21,** 51 (1970).
53. D. A. Dowden, *Colloq. sobre Quim. Fis. Processos en Superficias, Madrid* p. 177, 1965.
54. T. W. Hickmott and G. Ehrlich, *J. Chem. Phys.* **24,** 1263 (1956).
55. P. A. Redhead, *Trans. Faraday Soc.* **57,** 641 (1961).
56. G. Ehrlich, *J. Chem. Phys.* **34,** 39 (1961).
57. J. W. Coburn, *Surface Sci.* **11,** 61 (1968).

58. D. A. Degras, *J. Chim. Phys.* 406 (1967).
59. D. A. Degras and J. Lecante, *Nuova Cimento Suppl.* **5,** 598 (1967).
60. A. G. Jackson and M. P. Hooker, *Surface Sci.* **6,** 297 (1967).
61. R. P. Eischens and W. A. Pliskin, *Advan. Catal.* **10,** 1 (1958).
62. G. Blyholder, *J. Phys. Chem.* **68,** 2772 (1964).
63. G. Blyholder, *Proc. Int. Congr. Catal. 3rd, 1964* p. 657. North-Holland Publ., Amsterdam, 1965.
64. C. W. Garland, R. C. Lord, and P. F. Troiano, *J. Phys. Chem.* **69,** 1188, 1195 (1965).
65. A. C. Yang and C. W. Garland, *J. Phys. Chem.* **61,** 1504 (1957).
66. K. C. Lin, J. D. Witt, and R. M. Mammaker, *J. Chem. Phys.* **55,** 1148 (1971).
67. A. G. Whitney, *J. Catal.* **25,** 176 (1972).
68. R. L. Banks and G. C. Bailey, *Ind. Eng. Chem. Prod. Res. Develop.* **3,** 170 (1963).
69. A. Clark and C. Cook, *J. Catal.* **15,** 420 (1969).
70. J. C. Mol, J. A. Moulijn, and C. Boelhouwer, *Chem. Commun.* **633** (1968).
71. G. S. Lewandos and R. Pettit, *Tetrahedron Lett.* No. 11, 789 (1971).
72. R. Hoffmann and R. B. Woodward, *J. Amer. Chem. Soc.* **87,** 2046 (1965).
73. C. O. Nenitzescu, C. Huch, and A. Huch, *Angew. Chem.* **68,** 438 (1956).
74. F. Eirich and H. Mark, *J. Colloid Sci.* **11,** 748 (1956).
75. G. Bier, *Kunstoffe* **48,** 354 (1958).
76. G. Natta, *J. Inorg. Nucl. Chem.* **8,** 589 (1958).
77. H. Uelzmann, *J. Polym. Sci.* **32,** 457 (1958).
78. F. Patat and H. Sinn, *Angew. Chem.* **70,** 496 (1958).
79. J. Boor, Jr., *J. Polym. Sci.* **C1,** 257 (1963).
80. A. Gumboldt and H. Schmidt, *Chem. Z.* **83,** 636 (1959).
81. P. H. De Bruyn, *Chem. Weekblad.* **56,** 161 (1960).
82. P. Cossee, *J. Catal.* **3,** 80 (1964).
83. E. J. Arlman, *J. Catal.* **3,** 89 (1964).
84. E. J. Arlman and P. Cossee, *J. Catal.* **3,** 99 (1964).
85. P. Cossee, P. Ros, and J. H. Schachtschneider, *Int. Congr. Catal., 4th, Moscow* Paper 14 (1968).
86. J. Chatt and L. A. Duncanson, *J. Chem. Soc.* 2939 (1953).
87. D. D. Eley, "Cationic Polymerization and Related Complexes" (P. H. Plesch, ed.), p. 1. Academic Press, New York, 1953.
88. A. Clark and J. R. Harris, *J. Catal.* **21,** 179 (1971).
89. R. A. Harman and H. Eyring, *J. Chem. Phys.* **10,** 557 (1942).
90. J. R. Anderson and C. Kemball, *Proc. Roy. Soc. (London) Ser. A* **226,** 472 (1954).
91. F. G. Gault and C. Kemball, *Trans. Faraday Soc.* **57,** 1781 (1961).
92. J. J. Rooney, *Chem. Brit.* **2,** 242 (1966).
93. E. P. Parry, *J. Catal.* **2,** 371 (1963).
94. J. E. Mapes and R. P. Eischens, *J. Phys. Chem.* **58,** 1059 (1954).
95. D. E. Nicholson, *Nature (London)* **186,** 630 (1960).
96. M. R. Basila and T. R. Kanter, *J. Phys. Chem.* **70,** 1681 (1966).
97. M. R. Basila, T. R. Kanter, and K. H. Rhee, *J. Phys. Chem.* **68,** 3197 (1964).
98. M. R. Basila and T. R. Kanter, *J. Phys. Chem.* **71,** 467 (1967).
99. J. A. Cusumano and M. J. D. Low, *J. Catal.* **23,** 214 (1971).

III

Analogies with Simple Chemical Bonds—The Ionic Bond

In this chapter, we shall deal chiefly with ionic chemisorptive bonds on *metal* surfaces. Ionic chemisorption on nonmetallic surfaces has certainly received a great deal of attention, but results are interpreted mostly in terms of band theory (see Chapters V and IX). Interpretations of ionic chemisorption on metal surfaces are often based on the *free-electron theory* of metals, particularly electrical conductivity. Before going into the results of chemisorption studies, we shall describe the theory and facets of it pertinent to surface problems, including the Fermi level, the potential in a metal, and the work function. Further, we shall mention surface potentials and surface dipoles. Although the introduction of the free-electron theory is a concession to the collective-electron approach, the analogy with localized chemical bonding is preserved in the simple theories of this chapter. The theories are based on the use of experimental values of ionization potentials, electron affinities, work functions, and mirror-image potentials, and in a few instances, the use of quantum-mechanical procedures. They show only fair agreement with experiment and work best for chemisorbed atoms of low ionization potential, such as alkali metal atoms. In the final section, qualitative descriptions of ionic adsorption complexes will be given.

3.1 Electronic Surface Phenomena†

An important quantity in the study of systems of particles is the energy distribution function. The expression for the distribution function depends on the types of particles. We define the distribution function for a system as the mean number of particles in a given energy state, ϵ_j. In a *non-quantized* system of *distinguishable* or *indistinguishable* particles of a classical ideal gas, for example, the distribution function $\bar{n}_j$ is

$$\bar{n}_j = 1/\exp[(\epsilon_j - \mu)/kT], \tag{1}$$

provided the number of energy states is far greater than the number of molecules in the system, a condition almost invariably met under usual circumstances. The quantity μ is the chemical potential of the system. Statistics of this form are called *classical* or *Maxwell–Boltzmann*. When the particles are considered *indistinguishable* and *quantized*, *Bose–Einstein* statistics apply, and the distribution function is

$$\bar{n}_j = 1/\{\exp[(\epsilon_j - \mu)/kT] - 1\}. \tag{2}$$

Particles which behave according to Bose–Einstein statistics are, for example, phonons, photons, magnons; and ^{4}He at low temperatures. When the particles are *indistinguishable, quantized*, and restricted to *one particle per energy state, Fermi–Dirac* statistics apply, and the distribution function is

$$\bar{n}_j = 1/\{\exp[(\epsilon_j - \mu)/kT] + 1\}. \tag{3}$$

Electrons in metals, nucleons, and ^{3}He gas at low temperature behave according to Fermi–Dirac statistics. The restriction to one particle per energy state is an expression of the Pauli exclusion principle and means, in the case of electrons, that no two may have all quantum numbers the same; they must differ at least in spin ($\pm\frac{1}{2}$).

The free-electron theory of metals is based on Fermi–Dirac statistics. A metal is pictured as a box of mobile valence electrons to which the quantum mechanics of a particle-in-a-box are applied. Since only one electron is allowed per energy state, electrons will fill the states up to some maximum level. These are translational states and form essentially a continuum of levels. Each energy level is degenerate, containing two energy states differing by values of electron spin ($\pm\frac{1}{2}$). At absolute zero, the maximum level will be sharply defined; below this level all energy states will be filled and above it all will be empty. At higher temperatures, the cutoff will not be so sharp, with some electrons above the "maximum"

† Refs. [1–3].

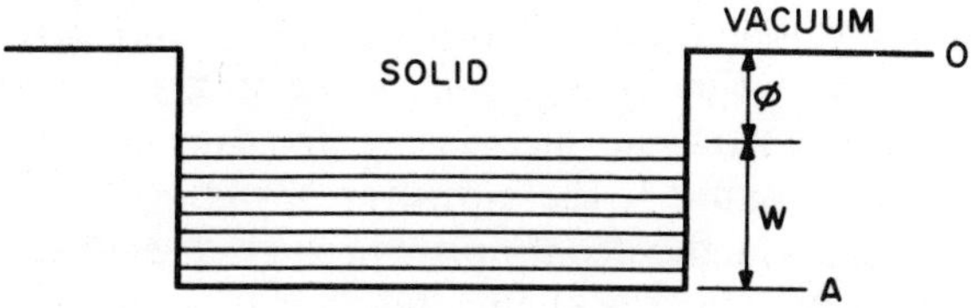

Fig. 3.1. Electron levels in a metal at low temperature.

level and an equal number of vacancies below it. It turns out that the maximum level is given by μ, the chemical potential of the system, Eq. (3), and it is called the Fermi level, E_F. Inspection of Eq. (3) reveals that the mean number of electrons in a particular state never exceeds unity, which agrees with the restriction of one electron per energy state. Furthermore, at absolute zero when $\epsilon_j > \mu$, the value of $\bar{n}_j$ is zero, and when $\epsilon_j < \mu$, it is unity, showing the sharp cutoff at μ ($= E_F$). At higher temperatures, $\bar{n}_j$ may take on values between zero and unity for the conditions $\epsilon_j > \mu$ or $\epsilon_j < \mu$, showing the fuzziness which exists around the high-energy states. But at all temperatures, the value of $\bar{n}_j$ when $\epsilon_j = \mu$ is $\frac{1}{2}$, and this becomes the general definition of the Fermi level: that energy level at which the mean number of electrons per energy state is one half; in other words, the probability that there will be an electron in an energy state at the Fermi level is one half. The potential energy of an electron in a metal is the work required to take it from its energy level in the metal to infinity. Obviously its value depends on what level it occupies in the metal, which may vary from the ground state to the Fermi level. It is the sum of two factors: (1) the energy to bring an electron to the Fermi level, W; and (2) the energy to bring it from the Fermi level out of the metal to infinity, ϕ.

$$\mathrm{PE} = -(\phi + W) \tag{4}$$

where the minus sign is the convention. For ionic chemisorption, the important quantity is ϕ, the work function, giving the energy required to bring an electron from the top energy level—the Fermi level—to infinity. The physical relationships among PE, ϕ, and W are shown in Fig. 3.1 at absolute zero. At higher temperatures, ϕ varies slightly as the boundary between occupied and unoccupied states becomes diffuse.

As a layer of charged chemisorbed particles builds up on a metal surface, it will facilitate or hinder the expulsion of electrons from the metal, depending on the charge. Thus, the work function will gradually increase as the coverage by negatively charged adsorbed particles increases, and decrease with positively charged particles. Adsorption may not cause charge separation to the point of creating ions; the separation may be partial with the formation of dipoles. Depending on their orientation, these dipoles will also cause an increase or decrease in the work function with coverage. In Section 2.1, the measurement of surface potentials and the estimation of surface dipoles from them were discussed. We note that many observations, such as the change in work function, indicate transfer of electrical charge on adsorption. But in the case of metals, it is very difficult to assign an exact electrical charge to an adsorbed particle.

3.2 Simple Theories of the Ionic Chemisorptive Bond

Rough estimates of the change in energy associated with ionic chemisorption of atom A on metal M

$$\mathrm{M} + \mathrm{A} \rightarrow \mathrm{M}^{-}\mathrm{A}^{+} \tag{5}$$

may be made by breaking down the energy mechanism as follows.

1. Remove an electron from the highest occupied level of an isolated atom of A to infinity $(-I)$.
2. Transfer the electron to the Fermi surface of the metal M (ϕ).
3. Bring A^+ to its equilibrium distance z^* above the surface of the metal $(e^2/4z^*)$.

The heat of adsorption $(Q_0 = -\Delta H)$ per gram atom of A at zero coverage is then given by

$$Q_0 = [-I + \phi + (e^2/4z^*)]N \tag{6}$$

where I and ϕ are the ionization potential of A and the work function of M, respectively, and N is Avogadro's number. The last term on the right is the energy of attraction of the ion A^+ to the metal surface and has been estimated to be equivalent to the electrostatic attraction of a charged particle above the surface of the metal to a charge of the same magnitude but opposite in sign the same distance below the surface. It is known as the classical mirror-image energy, and the equilibrium distance of separation of the charged particle and its negative counterpart is $2z^*$. A similar expression for the negative chemisorbed ion

$$\mathrm{M} + \mathrm{A} \rightarrow \mathrm{M}^{+}\mathrm{A}^{-} \tag{7}$$

can be written

$$Q_0 = [-\phi + A_0 + (e^2/4z^*)]N \tag{8}$$

where A_0 is the electron affinity of adsorbate A. Note that $+\phi$ and $-\phi$

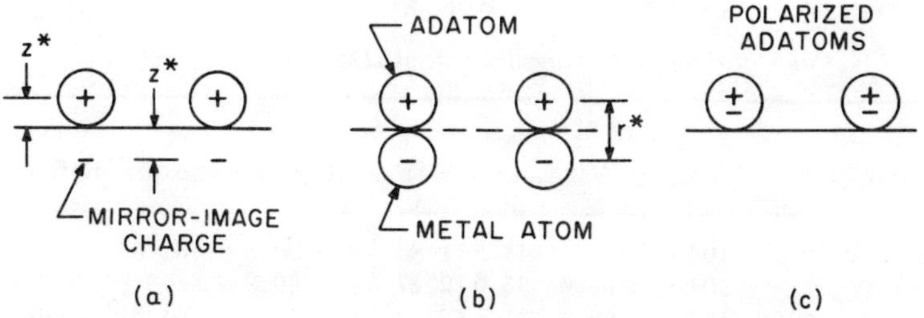

Fig. 3.2. (a) Adsorption of an ion; the negative charges are the electrical mirror images of the positive charges on the adions. (b) Ionic adsorption on individual metal atoms. (c) Van der Waals adsorption with adatoms polarized by a surface field.

for the metal correspond to the electron affinity and ionization potential of an isolated atom, for the electronic structure of a metal is not changed significantly by the gain or loss of an electron. If the adsorption is dissociative, as in the case of hydrogen and other diatomic molecules, then $-\frac{1}{2}D_{AA}$, the heat of dissociation of A_2 per gram atom, must be added to the right-hand side of Eqs. (6) and (8).

In the foregoing procedure, the surface of the metal is considered to be an infinite plane, and z^* is assumed to be the radius of the adsorbed ion, as shown in Fig. 3.2a. If we consider electronic interaction and charge transfer between an adsorbate atom and a single metal surface atom, the situation would be depicted as in Fig. 3.2b. Van der Waals adsorption with polarization of the adatoms, forming dipoles, is illustrated in Fig. 3.2c. In addition to the question of the proper value for z^*, other criticisms of the validity of Eqs. (6) and (8) can be made. For instance, the classical mirror-image law is not adequate at distances comparable with lattice spacing; also, short-range quantum-mechanical forces of attraction and repulsion have been ignored.

Values of $N\phi$ lie between 90 and 140 kcal/gm atom for metals other than the alkali and alkaline earth metals; values of NI for oxygen, nitrogen, and hydrogen atoms are 312.0, 333.5, and 312.0 kcal/gm atom, and the classical mirror-image energy $Ne^2/4z^*$ is in the neighborhood of 50 kcal/gm atom. Using values in this range for $N\phi$, NI, and the image energy in conjunction with the dissociation energies of oxygen, nitrogen, and hydrogen (117, 225, and 103 kcal/mole, respectively), we calculate extremely high endothermic values. For example, the heat of adsorption of hydrogen on tungsten

$$W + \frac{1}{2}H_2 \rightarrow W^-H^+$$

is approximately -210 kcal/gm atom. Endothermic processes of this magnitude are highly improbable. Similar conclusions are reached for

TABLE 3.1

Calculated and Experimental Initial Heats of Chemisorption[a]

System	$N\phi$	NI	$Ne^2/4z^*$	Q_0 (calc.)	Q_0 (exp.)	Reference
Na on W	104	118	44.5 (1.83 Å)[b]	30.5	32.0	[5]
K on W	104	99.6	35.9 (2.27 Å)	40.3	—	
Cs on W	104	89.4	31.1 (2.62 Å)	45.7	64.0	[6]

[a] Values in kcal/gm atom.
[b] Values in parentheses, z^*.

adsorption as the negative ion. We may conclude that for oxygen, nitrogen, and hydrogen a purely ionic bond will not form. But the adsorption of alkali metals, whose ionization potentials are low, appears to proceed with the formation of positively charged adsorbed ions, as may be seen in Table 3.1, which is from Trapnell and Hayward [4].

In the previous section, we mentioned the change in work function with buildup of charge on a metal surface. Several workers have suggested that the fall in heat of adsorption with coverage can be attributed to $\Delta\phi$, the change in work function [5–7]. The problem has been discussed in detail by Higuchi *et al.* [8]. They start with the Helmholtz equation

$$V = 4\pi B\theta ez^*/\epsilon \tag{9}$$

where V, the contact potential, is the difference in potential between the adsorbed layer and the metal surface, and therefore is equal to $\Delta\phi$, the change in work function; B is the number of adsorption sites per unit area; θ the fraction of sites covered; ϵ the dielectric constant of the medium in the electric double layer formed by the adsorbed film; e the electric charge; and z^* the distance shown in Fig. 3.2a for the adsorption of an ion with its mirror image. Equation (9) may also be put into the appropriate form for interaction of an adatom with a single metal atom (Fig. 3.2b) by replacing z^* by the distance r^*, the sum of the radii of the adatom and a surface metal atom, and er^* by μ, the dipole moment. Setting $\epsilon = 1$, we obtain

$$V = 2\pi B\mu\theta. \tag{10}$$

The contact potential V arises from electron transfer. In desorption, the electron is transferred back to the original position through the potential difference $V = \Delta\phi$. If the adatom is positive, that is, an electron is transferred to the metal, the potential as it increases with coverage favors more and more the transfer of the electron back to the adatom. If the adatom is negative, the potential as it increases with coverage favors more and more the transfer of the electron back to the metal. Thus, in both cases, the heat of desorption decreases with coverage and is given approximately by

$$\Delta Q = N\ \Delta\phi. \tag{11}$$

When homonuclear diatomic molecules chemisorb as atoms and desorb as molecules, Eq. (11) becomes

$$\Delta Q = 2N\ \Delta\phi. \tag{12}$$

Neither Eq. (11) nor Eq. (12) has been tested sufficiently to allow conclusions about its validity for ionic chemisorption.

3.3 Quantum-Mechanical Approaches

Our first example, a proposal by Higuchi *et al.* [8], works in principle for bonds ranging from purely covalent to purely ionic. The bonds are considered strictly localized, consisting of the adsorbed atom A and a surface atom M. The wave function of the bond is written

$$\psi = C_i\psi_i + C_c\psi_c \tag{13}$$

where ψ_i and ψ_c are the wave functions of the ideal ionic and covalent bonds, respectively, and C_i and C_c are constants. Equation (13) is a linear combination of ionic and covalent wave functions, each of which could be broken down into a linear combination of atomic orbitals (see LCAO–MO method, Section 2.3), but that is not necessary in this example. Just as in Section 2.3, the wave equation $H\psi = E\psi$ is put in the form $(\int\psi H\psi\, d\tau)/\int\psi^2\, d\tau = E$, and the energy E of the chemisorptive bond is minimized with respect to C_i and C_c by the usual variational procedure. The following expression is obtained.

$$1/C_i^2 = 1 + (E - H_{ii})/(E - H_{cc}) \tag{14}$$

where $H_{ii} = \int\psi_i H\psi_i\, d\tau$ and $H_{cc} = \int\psi_c H\psi_c\, d\tau$. The fractional ionic character of the bond is given by $C_i^2 = 1 - C_c^2$, provided that $\int\psi_i^2\, d\tau$ and $\int\psi_c^2\, d\tau$ are normalized. From Eq. (14) we find, for a purely ionic bond $(C_i^2 = 1)$, that $E = H_{ii}$. Similarly, for a purely covalent bond $(C_c^2 = 1)$, we find that $E = H_{cc}$. Now the problem is to develop ways of determining C_i, H_{ii}, and H_{cc}, so that the bond energy E may be obtained from Eq. (14). Higuchi *et al.* [8] approximate H_{ii} by

$$H_{ii} = [A_0 - I + (8/9)(e^2/r^*)]N \tag{15}$$

where A_0 is the electron affinity of atom M and I the ionization potential of atom M. The last term in the brackets in Eq. (15) is the counterpart of the mirror-image term $e^2/4z^*$ in Eqs. (6) and (8), which was obtained on the assumption of a collective interaction of the metal with an adsorbate atom rather than interaction with a single metal atom. In the latter case, the energy u between a positive ion and a negative ion is assumed to be

$$u = -(e^2/r) + (B_r/r^9), \tag{16}$$

where the first term on the right is the electrostatic energy of attraction between two charges of opposite sign and the second term is the electron repulsion, proportional to the ninth power of the distance of separation r, and effective only at very small distances. When the energy u is at its minimum value, $(\partial u/\partial r) = 0$ at r^* and

$$u(r^*) = (8/9)(e^2/r^*). \tag{17}$$

A slight concession to the presence of a metal surface is made by assuming that r^* is the sum of the *metallic* radius of M and the covalent radius of A. For the completely covalent bond, Higuchi *et al.* follow Eley and Pauling (Section 2.1), assuming that

$$H_{cc} = \tfrac{1}{2}(D_{MM} + D_{AA}). \tag{18}$$

They calculate the fractional ionic character of the bond, C_i^2, from the expression

$$\mu = C_i^2 e r^* \tag{19}$$

where μ is the bond dipole moment and $C_i^2 e$ gives the effective charge, which is less than e, except when the bond is totally ionic. The bond moment μ is obtained from measurements of surface contact potentials as outlined in Section 2.1. Using this value of C_i^2 and values of H_{ii} and H_{cc} from Eqs. (15) and (18), the bond energy E is calculated from Eq. (14). The heat of adsorption per gram atom of a monatomic gas at zero coverage is

$$Q_0 = E \tag{20}$$

and per gram mole of a homonuclear diatomic gas A_2 is

$$Q_0 = 2E - D_{AA}. \tag{21}$$

For bonds such as Cs on W and Na on W, Higuchi *et al.* calculate values of $C_i^2 \approx 1$, that is, the bonds are purely ionic. But they find that bonds of Ba on W and Sr on W are ionic with a small amount of covalent character ($C_i^2 = 0.97$ for Ba and 0.67 for Sr). For diatomic gases such as oxygen, nitrogen, and hydrogen, the bonds are covalent with a small amount of ionic character ($C_i^2 = 0.02$–0.09). In strongly covalent systems, they find bond energies differing little from those obtained by the Eley–Stevenson method (Section 2.1), which could have been anticipated, since the approximations have much in common. Although the method of Higuchi *et al.* is more versatile, treating all degrees of bonds from pure covalent to pure ionic, it suffers from the same general weaknesses discussed under the Eley method. Other methods of estimating C_i^2 have been proposed and corrections for the perturbation of the metal–metal bond have been introduced [9]. Indeed, better agreement with experimental heats is obtained, but the methods are wholly empirical and do not lead to a better understanding of the nature of the chemisorptive bond.

The work of Gomer and associates [10–13] is broader in concept; it considers both localized and nonlocalized chemisorptive bonds, and ionic, covalent, and polar adsorption. They and others [14] have studied the difficult problem of the effective electronic charge q localized in the neighborhood of the adsorbate. For ionic adsorption,

$$q \approx \pm|e| \tag{22}$$

where $|e|$ is the absolute value of the charge on the electron. For neutral or covalent adsorption,

$$q \approx 0, \tag{23}$$

and for *metallic* adsorption,

$$0 < |q| < |e|. \tag{24}$$

Metallic adsorption is nonlocalized and involves the many-electron problem—all of the electrons of the metal plus the electron of the adsorbate.

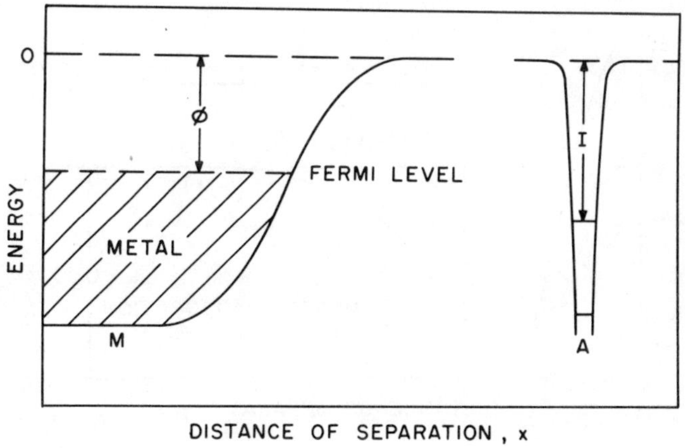

Fig. 3.3 Metal–adsorbate system when interaction is negligible.

Quantum-mechanical wave functions in this case extend throughout the metal and the adsorbate atom. No single electron state is localized around the adsorbed-atom position; the localized concentration of charge, q, is made up of an infinitely large number of electron states, each contributing an infinitesimal amount. Such bonds are considered polar in nature rather than purely ionic, and the degree of polarity depends on the proportionate total time that the adsorbed-atom level is occupied by the itinerant electrons. We will take up metallic adsorption in Chapter VIII.

First, consider the metal–adsorbate system when interaction is negligible, that is, when the distance of separation x is large. The situation is shown schematically in Fig. 3.3 for metal M and atom A. The metal may be described as a "potential box" of electrons obeying the free-electron theory, or as the conduction band separated by an energy gap from the valence band according to the band theory of metals; a differentiation between the two theories is not necessary for the objectives of this section.

As the distance of separation between the metal and atom A is decreased, we are interested in the effect on the electron levels of both members of the system. We will be primarily concerned with the case $M + A \rightarrow M^- + A^+$, where A loses an electron to the metal at or below the Fermi level. When the interaction is small, $M^- + A^+$ lies above $M + A$ by $I - \phi$, assuming that $|I| > |\phi|$. Decreasing the distance of separation requires the introduction of a Coulombic or ionic interaction term (assuming A has developed a charge), giving $I - \phi - |V_i(x)|$, where $|V_i(x)|$ may be approximated by the image potential $|(e^2/4x)|$. Further decrease in the distance of separation leads to quantum-mechanical exchange interaction with the metal, and a covalent term $|V_c(x)|$ must

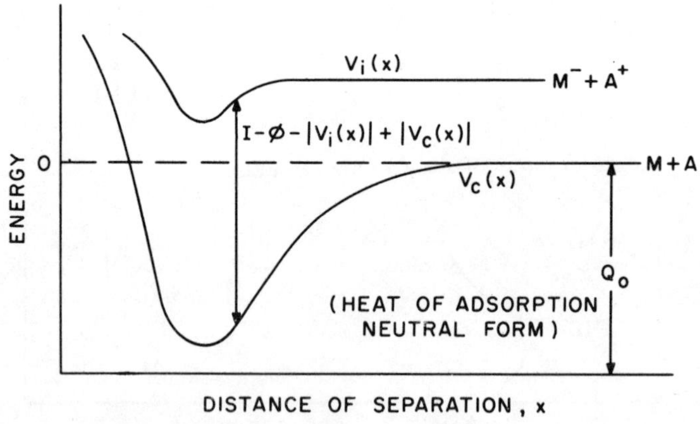

Fig. 3.4. Schematic potential energy diagrams for adsorption by localized bonding. Electronic term values $V_c(x)$ and $V_i(x)$ for covalent $M + A$ and ionic $M^- + A^+$ states are shown as a function of metal–adsorbate distance.

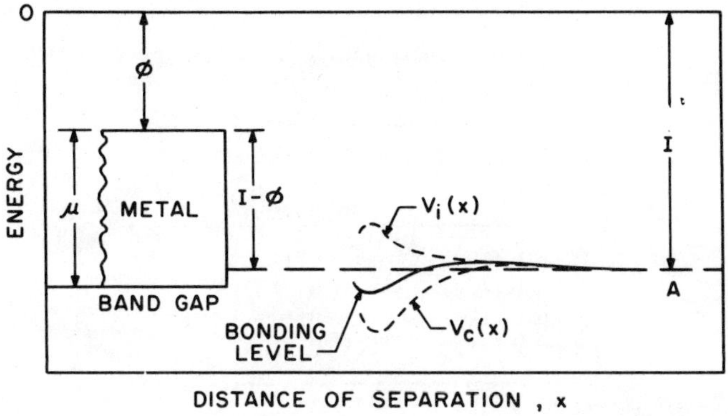

Fig. 3.5. Schematic electronic potential energy diagram showing the location of the bonding level as a function of M–A distance. Note that bonding level falls below lower edge of conduction band, so that bonding is by localized electrons.

then be included. At such small separation, $|V_i(x)|$ still exists, but can no longer be represented by the simple relative magnitudes of $V_i(x)$ and $V_c(x)$. If the electron level of A falls below the bottom of the conduction band (see Fig. 3.3), adsorption with localized states is possible, provided the electron level of A is not fully occupied. When the highest level of A is occupied—that is, loss of an electron to the metal has not occurred, as in the adsorption of rare gases on metals—interaction will be weak. If the highest occupied level of A falls within the range of energies of the conduction band electrons, adsorption will be metallic in nature, a condition probably true for the adsorption of alkali and alkaline earth metal atoms on tungsten or molybdenum, which explains why these systems are polar, not necessarily purely ionic. When more than one particle is adsorbed, the reference particle is situated in a potential dipole layer formed by all the adparticles, changing the effective charge and the heat of adsorption. The concept of localized adsorption is illustrated schematically in Figs. 3.4 and 3.5. The first of these two figures shows the potential energy diagrams of the adsorbed atom as a function of distance of separation, and the second relates the interactions and bonding levels to the energy levels in the metal. Note that the bonding level falls below the bottom edge of the conduction band of the metal, so that bonding is localized.

3.4 Ionic Surface Complexes

Experimental studies indicate that hydrogen chemisorbed on transition metals possesses only a very small ionic character. For example, the work of Collins and Blott [15] on the dissociative adsorption of hydrogen on

TABLE 3.2

Dipole Moments of Hydrogen Chemisorbed on Tungsten Crystal Faces[a]

System	μ, Dipole moments (debyes)
H on W(100)	0.144
H on W(112)	0.276
H on W(110)	0.109–0.124
H on W(111)	0.110

[a] Ref. [15].

the crystallographic faces of tungsten shows (see Table 3.2) that the dipole moments, determined from surface potentials, are small. The order of magnitude of the charge on the dipoles is $0.1e$ and the separation about 0.01 Å. Their results agree with a model based on bonding to two prominent neighboring surface atoms of tungsten. Theoretical calculations are in accord with these results. Using an equation suggested by Pauling [16] for the fractional ionic character of surface bonds, $(1 - \exp(-0.25\mu^2))$, we find very low values. The simple theories of Section 3.2 indicate strongly that significant ionic bonding is unlikely. Further, Grimley [17] extends the simple ionic equation (8) to include a Coulombic term, which becomes the mirror-image potential at large distances, and a quantum-mechanical exchange interaction term, and shows that chemisorption as H^- cannot be justified for metals.

In the case of dissociative chemisorption of oxygen, the question of the polarity of the surface bond on transition metals has not been settled unequivocally. Originally, the relatively small surface potential was interpreted [18] to mean that oxygen adsorption was essentially atomic. But MacRae [19] reasons that the surface potential data may not preclude a considerable charge transfer to the adsorbed oxygen. Park and Farnsworth on the basis of photoelectric data [20] favor atomic adsorption. For molybdenum and tungsten, the saturation values for oxygen chemisorption are sufficiently low that there is little doubt that oxygen take-up is confined to the surface. But on cobalt and nickel the saturation values at room temperature are high enough to raise doubts whether all of the oxygen is on the surface. Brennan and Graham [21] present evidence for subsurface oxide formation in the cobalt and nickel systems.

Nitrogen is adsorbed in three different forms on tungsten [22]: strong (β), intermediate (α), and weak (γ). The specificity of the various crystallographic planes is striking. On the (110) plane, only weak (γ) adsorption is observed at temperatures below 150°K and no adsorption occurs at 300°K. Under the same conditions, other planes show considerably stronger chemisorptive power (see Table 3.3). In all cases, dipole moments are low, indicating very small ionic character.

Studies of the adsorption of carbon monoxide on tungsten planes by flash filament desorption [23–28], field emission [12, 29], and electron impact desorption [29] give evidence for the existence of at least three types of chemisorption states, differing not only in their heats of adsorption and electron desorption cross sections, but also in the sign of their dipole moments. Again, the dipole moments are small, indicating low ionic character.

The ionic characters of alkali and alkaline earth atoms chemisorbed on transition metals are far greater than those of the simple diatomic gases discussed above. Yet Utsugi and Gomer [11], considering the Cs–W system, prefer to call the adsorption bond polar rather than completely ionic. Their reason is that with a broadening of the Cs s level the band has some metallic character, which results in part of the broadened level's lying within the conduction band below the Fermi level. In other words, the image potential of Cs^+–W^- cannot raise the Cs s band totally above the Fermi level except at Cs surface separations so small that the image potential no longer describes the interaction and a repulsive term dominates. Levine and Gyftopoulos [30] believe that Cs atoms are adsorbed on W as a single species, each Cs–W bond having covalent and ionic

TABLE 3.3

Nitrogen Adsorption on Individual Tungsten Planes[a]

Plane	T (°K)	State	Desorption energy (kcal/mole)	Association	Dipole (D)
(110)	110	γ	~9	Atoms	>0.04
(100)	110	γ^+	~10	Molecules	>0
		γ^-	~11	Molecules	<0
	300	β	75	Atoms	0.4
(111)	110	γ	~9	Molecules	0.1
	220	α	16	Molecules	<−1
	300	$\beta_1 + \beta_2$	~75	Atoms	−0.02

[a] Ref. [22]

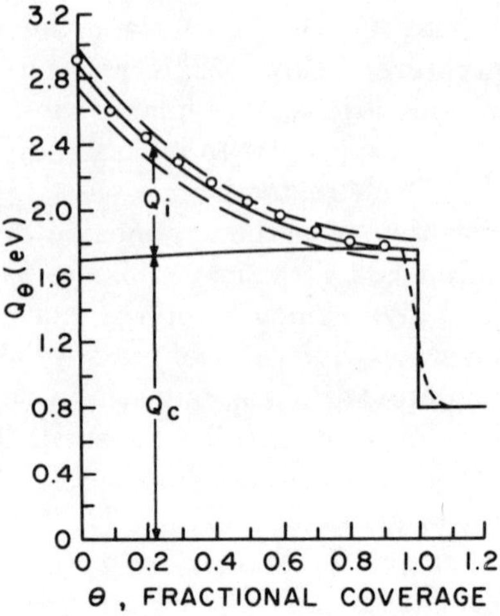

Fig. 3.6. Comparison of theoretical (solid curve) atom desorption energy Q_θ for Cs on W with Taylor–Langmuir data (circles). θ is the film coverage expressed in monolayers. The 3% uncertainty limits are shown by dashed lines, and the ionic and covalent contributions Q_i and Q_c are also depicted. The dotted line for $\theta \approx 1$ schematically represents Q_θ when the second layer formation is taken into account [After Levine and Gyftopoulos [30]. By permission of North-Holland Publ. Company.]

parts and a total bond energy

$$Q_\theta = Q_i + Q_c \tag{25}$$

where Q_θ is the bond energy at fractional coverage θ and Q_i and Q_c are the contributions to the bond energy of ionic and covalent parts, not to be construed as the sum of a completely ionic and a completely covalent bond. They calculate the ionic part as follows for a desorption process:

1. Remove the adsorbed particle with fractional charge F to infinity, leaving an equal and opposite charge in the metal;
2. Remove the fractional electron charge from the surface to infinity;
3. Combine the fractional ion with the fractional electron charge to produce a neutral atom.

Summing these three steps, they obtain

$$Q_i = F\phi(1 + \delta), \tag{26}$$

where ϕ is the work function of the metal, and

$$\delta = F(e^2/r - I)/\phi \tag{27}$$

with e^2/r the electrostatic potential and I the ionization potential of the adsorbed atom. The stepwise mechanism described above is comparable to the mechanism described in Section 3.2, which was used in obtaining Eq. (6), except that fractional charges are involved in the present case. An expression for calculating Q_c is obtained by combining contributions from the adsorbate and the metal, each assumed to be proportional to the heat of sublimation, the square of the angular strength of the valence orbital, and the valence charge. Calculated values of $Q_\theta = Q_i + Q_c$ are compared in Fig. 3.6 with experimental values determined as a function of surface coverage from data of Taylor and Langmuir [31]. Note that the ionic part of Q_i decreases to zero at $\theta = 1$, while the covalent part Q_c changes little with coverage. The agreement with experiment is good for this case. Similar results have been obtained by Schmidt and Gomer [13] for potassium chemisorbed on individual crystallographic planes of tungsten. Adsorption appears to be polar, possibly wholly ionic on the (110) plane at low coverage. With increasing coverage, adsorbate charge decreases markedly but smoothly on all planes.

The controversy over localized versus nonlocalized (metallic) bonds continues to rage. Sachtler [32], a strong proponent of localized adsorption, contends that surface compound formation by chemisorption "demetallizes" the surface metal atoms as they form nonmetallic compounds and lose their collective properties. Yet he concedes that transition metal atoms with a large number of stable valence states, Mo for example, may continue to contribute appreciably to collective properties, even though one or two electrons per metal atom are used for chemical bonding. Furthermore, for these metals, partial demetallization of surface atoms may affect different collective properties. The concepts which we have reviewed in this section would seem to indicate that very small changes in energy levels can cause them to be located inside or outside the conduction band energies, and thus determine whether adsorption is metallic or localized. Chemisorption systems are highly specific and generalizations are dangerous.

REFERENCES

1. A. Clark, "The Theory of Adsorption and Catalysis," p. 193. Academic Press, New York, 1970.
2. T. L. Hill, "An Introduction to Statistical Thermodynamics," Chapter 22. Addison-Wesley, Reading, Massachusetts, 1960.
3. R. A. Levy, "Principles of Solid State Physics," Chapter 7. Academic Press, New York, 1968.
4. B. M. W. Trapnell and D. O. Hayward, "Chemisorption" (W. E. Garner, ed.), p. 210. Butterworth, London and Washington, D.C., 1964.
5. M. Boudart, *J. Amer. Chem. Soc.* **74**, 3556 (1952).

6. J. C. Mignolet, *Bull. Soc. Chim. Belg.* **64,** 126 (1955).
7. J. H. De Boer and C. F. Veenemans, *Physica* **1,** 960 (1934).
8. I. Higuchi, T. Ree, and H. Eyring, *J. Amer. Chem. Soc.* **79,** 1330 (1957).
9. L. Manes and E. Molinari, *Z. Phys. Chem.* **39,** 104 (1963).
10. L. W. Swanson and R. Gomer, *J. Chem. Phys.* **38,** 1613 (1963).
11. H. Utsugi and R. Gomer, *J. Chem. Phys.* **37,** 1720 (1962).
12. L. W. Swanson and R. Gomer, *J. Chem. Phys.* **39,** 2918 (1963).
13. L. D. Schmidt and R. Gomer, *J. Chem. Phys.* **45,** 1605 (1966).
14. A. J. Bennett and L. M. Falikov, *Phys. Rev.* **151,** 512 (1966).
15. R. A. Collins and B. H. Blott, *J. Appl. Phys.* **40,** 5390 (1969).
16. L. Pauling, "The Nature of the Chemical Bond." Cornell Univ. Press, New York, 1945.
17. T. B. Grimley, "Chemisorption" (W. E. Garner, ed.). Butterworth, London and Washington D.C., 1957.
18. P. M. Gundry and F. C. Tompkins, *Quart. Rev.* **14,** 257 (1960).
19. A. U. MacRae, *Surface Sci.* **1,** 319 (1964).
20. R. L. Park and H. E. Farnsworth, *Surface Sci.* **3,** 287 (1965).
21. D. Brennan and M. J. Graham, *Discuss. Faraday Soc.* **41,** 95 (1966).
22. G. Ehrlich, *Discuss. Faraday Soc.* **41,** 7 (1966).
23. G. Ehrlich, *J. Chem. Phys.* **36,** 1171 (1962).
24. P. A. Redhead, *Trans. Faraday Soc.* **57,** 641 (1961).
25. L. J. Rigby, *Can. J. Phys.* **42,** 1256 (1964).
26. J. Eisinger, *J. Chem. Phys.* **27,** 1206 (1957).
27. F. Gavrilyuk and G. Medvedev, *Fiz. Tverd. Tela* **4,** 2372 (1962) [*English transl.*: *Sov. Phys.-Solid State* **4,** 1737 (1963)].
28. T. E. Madey, J. T. Yates, and R. C. Stern, *J. Chem. Phys.* **42,** 1372 (1965).
29. D. Menzel and R. Gomer, *J. Chem. Phys.* **41,** 3329 (1964).
30. J. D. Levine and E. P. Gyftopoulos, *Surface Sci.* **1,** 171 (1964).
31. J. B. Taylor and I. Langmuir, *Phys. Rev.* **44,** 423 (1944).
32. W. M. H. Sachtler, *Surface Sci.* **22,** 468 (1970).

IV

Chemisorption and the d Band

Like all theories at the outset, the d-band theory of chemisorption and catalysis on transition metals was attractive, uncomplicated, and a potential answer to perplexing problems. Parallel changes in d-band vacancies and chemisorptive and catalytic power, demonstrated with metal alloys, led simply to the idea that electrons of the adsorbate entered "holes" in the d band and paired with metal d electrons to form predominantly covalent bonds. In time the complexities of the theory and exceptions to it grew. The theory was never truly quantitative; it depended chiefly on parallelisms between chemisorption or catalysis and changes in metal-alloy composition which were presumed to alter d-band vacancies in a known pattern. Indeed, quantitative theories are difficult to achieve in chemisorption and catalysis because the techniques for direct study of surfaces are still in their infancy. No one can say with certainty that d-band vacancies play no role in chemisorption. The melee of particles and the composition of forces within the confines of a metal deny such absolute statements about individual factors. All that can be said is that whatever the role, major or minor, we do not understand its full complexity yet. A consideration of the theory and the experimental work behind it are instructive.

We begin in the first section with a qualitative picture of the band theory of metals—how discrete energy levels broaden in solids, the overlapping of bands, and the specific nature of transition metals. In the next section, we review the experimental work, especially on metal alloys, which has been interpreted in terms of d-band vacancies. A discussion of various theoretical aspects follows; and, finally, we attempt a summing-up of the status of the subject with comments on the question of localized versus nonlocalized adsorption, and on alternative explanations of chemisorption on transition metals.

4.1 The Band Theory of Metals†

In Section 3.1, we discussed the simplest theory of collective electrons—the free-electron theory—in which the metal is pictured as a potential box containing the valence electrons in unhindered motion, and it is assumed that there are no repulsive forces between electrons. Another collective approach is the band theory, developed initially by Bloch [6], who introduced molecular orbital theory for this purpose, a theory later applied successfully to free molecules. In molecular orbital theory, each electron is assumed to move in a periodic potential field that extends over all atoms, and an approximate solution of the wave equation is described in terms of linear combinations of atomic orbitals. If the atomic orbital of any one atom overlaps significantly with more than one of its neighbors, then the molecular orbitals will be delocalized. In a molecule containing a finite number of atoms—benzene, for example—a p atomic orbital of each carbon atom overlaps equally in π bonding with both of its neighbors, resulting in molecular orbitals that extend over all six carbon atoms. Nothing prevents extension to an infinite array of atoms. An alkali metal, for example, has one valence electron in an s orbital which overlaps considerably with all 14 neighbors. We would expect complete delocalization of s electrons. But in one important respect, infinite arrays differ from the finite arrays of ordinary molecules. To see this, we start with a single atom and note the way that allowed energies of the electrons change as we build up the infinite crystal by successive additions of one more atom. For the moment, consider a single atomic orbital—the s orbital of lithium, for example. When a second atom B is added to the first one A, the atomic orbital of A, ψ_A, interacts with an identical orbital of B, ψ_B, giving two diatomic orbitals, $\psi_A \pm \psi_B$, each with its distinct energy. If a third atom is added, three molecular orbitals are obtained. To accommodate the electrons, the number of molecular orbitals must equal the number of atomic orbitals of the atoms infinitely separated, which, in this case, is one

† Refs. [1–5].

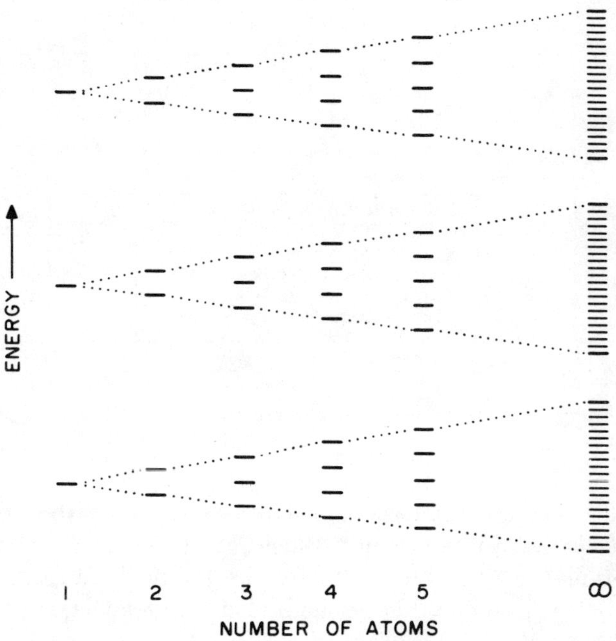

Fig. 4.1. The building up of energy bands by repeated addition of one more atom.

for each atom. We may continue the process indefinitely. Each new atom adds one more energy level to the array and slightly alters the values of the previous set. A diagram of the process in shown in Fig. 4.1. Observe that, in the limit, each atomic orbital gives rise to a band of energies. For each orbital of the original atom, there will be one band, and the total number of levels in a band is equal to the number of atoms in the metal. A monocrystalline unit of metal free of irregularities and cracks might contain 10^{10} atoms, in which case the energy levels would be so close that we are justified in calling them a continuum.

Quantitative calculations are difficult to make using an infinite linear combination of atomic orbitals. A more practical wave function was developed by Bloch [6], who started with the wave function for a free electron (potential energy constant),

$$\psi_k(x) = e^{ikx}, \tag{1}$$

where, for a linear array of atoms,

$$k = 2\pi l/Na, \qquad l = 0, 1, 2, \ldots, N-1 \tag{2}$$

and $Na = L$ is the length of the array, the product of N, the number of

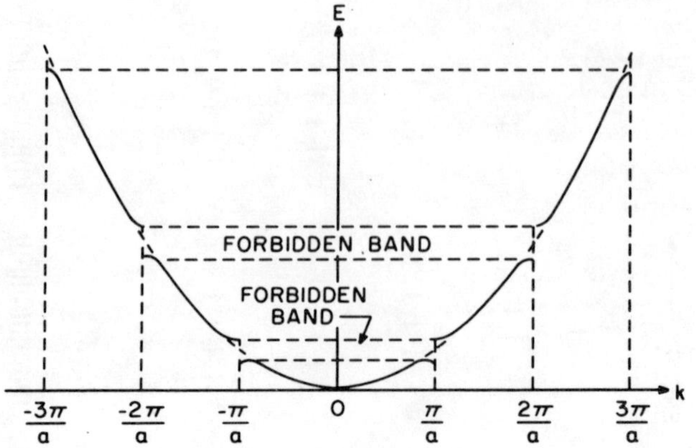

Fig. 4.2. Energy of an electron on a line of positive charges. Dashed curve represents free electron.

atoms, and a, the one-dimensional lattice constant. If the periodic field of the lattice is taken into account, Bloch reasoned that the wave function should be similar in type, but modulated by the field, and that the modulating function $u_k(x)$ should have the period of the lattice,

$$u_k(x + a) = u_k(x). \tag{3}$$

The function proposed by Bloch is

$$\psi_k(x) = e^{ikx}u_k(x) \tag{4}$$

and the proof that wave functions of an electron in a periodic field are of this type is known as *Bloch's theorem* in the theory of metals. Wigner and Seitz [7, 8] developed a method of solving the Schrödinger wave equation for Bloch-type functions. In Eq. (4), $u_k(x)$ becomes a constant for a free particle and the energy is found to be

$$E = \hbar^2k^2/2m, \tag{5}$$

that is, the energy is proportional to k^2. A plot of E versus k gives a parabola. The mathematics for obtaining the energy in the case of an electron appreciably influenced by a periodic field is complicated. We omit the details of evaluating $u_k(x)$ and the solution. But it is found that regardless of the shape of the potential curve, a small periodic perturbation such as $u_k(x)$ causes the energy levels to take sudden jumps at multiples of $k = \pm\pi/a$. As in the LCAO argument, the presence of energy bands and forbidden zones is clearly established (see Fig. 4.2). The three-dimensional problem is not essentially different from the one-dimensional problem, although it is considerably more complex.

We return to a consideration of the details of the band structure of metals. If the energy levels of the atoms are well separated or if the distance between nearest-neighbor atoms in the metal is great enough that the overlapping of atomic orbitals is small, the energy bands corresponding to the atomic orbitals will remain distinct, as in Fig. 4.1 and 4.2. But if the original atomic energy levels are close together or nearest-neighbor atoms are near enough for appreciable overlapping, the bands may lose their individuality and merge into one another. Thus bands may be the composite of two or more atomic orbitals. The energy levels of a metal are a function of the distance of separation of its atoms. At infinite separation, the energies are just those of the separate atomic orbitals. As the lattice constant decreases, interaction causes each separate level to split into a band, whose width increases steadily. Ultimately, the bands may cross, as shown in Fig. 4.3, calculated for metallic lithium. Obviously, band formation from atomic levels applies only to valence orbitals and higher ones. The orbitals of inner-shell electrons have little chance of interaction with neighboring orbitals, and thus these electrons remain essentially as localized as they do in the isolated atom.

We now summarize the band patterns of various groups of elements.

Alkali and noble metals

These elements possess a single valence electron and all should be metallic. Hydrogen is the only exception and forms a tightly bound diatomic molecule. In the others, there is considerable overlapping of bands. Since these metals have a single valence electron, electrical conductivity

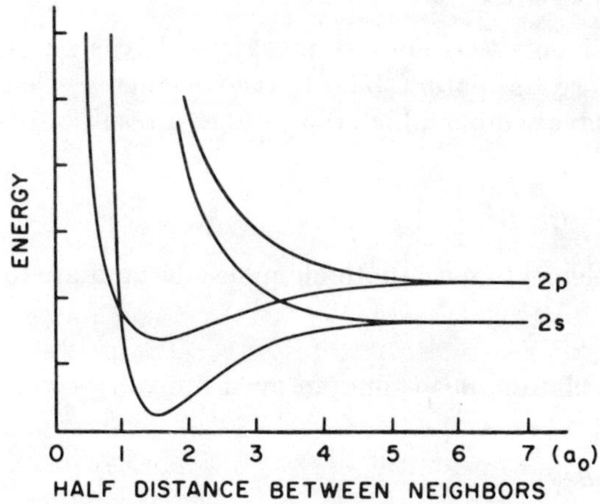

Fig. 4.3. Energy bands in metallic lithium as a function of distance between neighbors.

is assured, for electrons fill the bands from the bottom edge and the upper half of the band will be vacant.

Alkaline earth metals

These elements possess two valence electrons. They would be insulators if it were not for the fact that the valence band overlaps with higher unfilled bands, giving them true metallic qualities.

Group IIIA (B, Al, Ga, In, Tl)

These elements possess an odd number of electrons and are metallic. Boron is an exception which forms a covalent, nonmetallic solid.

Group IVA (C, Si, Ge, Sn, Pb)

These elements possess an even number of electrons and are insulators or poor conductors. The bands are well separated and full in the lower members. The heavier elements are more weakly bound and are semiconductors; the heaviest member, lead, has some overlap with empty bands and consequently, metallic properties.

Group VA (N, P, As, Sb, Bi)

These elements possess an odd number of electrons and might be expected to be metallic, but they tend to form covalent species with two atoms per cell. There is some band overlap in the higher members, which might be called semimetals.

Group VIA (O, S, Se, Te, Po)

The light elements (O_2 and S) possess widely separated filled bands and are, therefore, insulators. The heavier elements selenium, tellurium, and polonium have semimetallic properties as a result of slight overlapping of bands.

Group VIIA (F, Cl, Br, I, At)

All of these elements form diatomic molecules and are insulators.

Rare gases

These are insulators, possessing an even number of very tightly bound electrons.

Transition metals

For our present objectives, the transition metals are the most important. In the free atoms, energies of electrons in d states do not differ greatly

TABLE 4.1

Electrons in the 3d *and* 4s *Levels of Iron-Group Elements*

	Sc	Ti	V	Cr	Mn	Fe	Co	Ni
Number of 3d electrons	1	2	3	5	5	6	7	8
Number of 4s electrons	2	2	2	1	2	2	2	2

from the s or valence electrons. As examples, we shall consider the iron group (Sc, Ti, V, Cr, Mn, Fe, Co, and Ni), all of which have incomplete 3d subshells in addition to 4s electrons see (Table 4.1). In these metals, the 3d atomic states are neither broadened as much as those of the valence electrons nor as constricted as those of the inner shells. The d band will have an appreciable width at the interaction distance of the normal metal, but considerably less than that of the s electron. (We utilize the convenient concept of the widths of individual bands produced from specific atomic levels, even though the bands overlap.) Many of the properties of transition metals are explained on the assumption, with some corroboration by detailed calculations, that the 3d band is very narrow and overlaps the 4s band. Since the 3d atomic state is fivefold degenerate, the 3d band, in spite of its narrow width, contains more electrons than the 4s band, so that the density of states in the 3d band is considerably greater. Thus the energy distribution curve for electronic states in the 3d band is tall and narrow, while that of the 4s band is flat and wide. Nickel contains a total of ten electrons per atom in the 3d and 4s levels, and, since overlapping occurs, these electrons must be distributed between them (see Fig. 4.4d). Explicit calculations for nickel indicate that there are about 9.45 electrons in the d band, giving 0.55 hole, and 0.55 electron in the s band, $3d^{9.45}4s^{0.55}$. For iron, the distribution is $3d^{7.8}4s^{0.2}$.

We conclude this section with a few more remarks about the distribution of electronic states. The distribution function or density of states $N(E)$ is defined as the change in the number of energy states per unit volume per unit change in energy at energy E, that is,

$$N(E) = d\nu/dE \tag{6}$$

where ν is the number of states per unit volume with energies less than E. The expression $N(E)\,dE$ is the density of states with energies lying between E and $E + dE$. Examples of distributions are shown in Fig. 4.4. The first sketch (Fig. 4.4a) represents the density of states for a free-electron system corresponding to one valence electron per atom. The

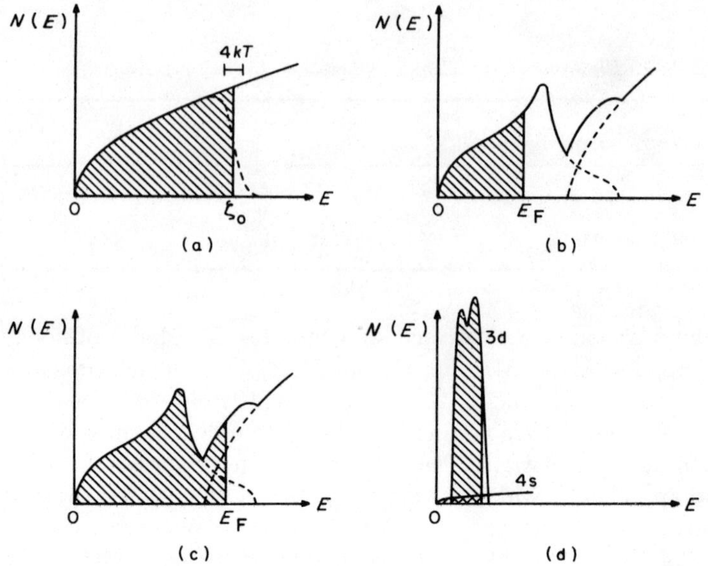

Fig. 4.4. Distribution curves. The shaded areas show the distribution of occupied states at absolute zero temperature. At higher temperatures, the cutoff in occupied states is less abrupt. (a) Free-electron system; (b) monovalent metal; (c) divalent metal; (d) transition metal, nickel.

shaded region gives the distribution of *occupied* states and its area is $\frac{1}{2}n$, where n is the total number of electrons and equal to twice the total number of occupied states ($n = 2\int_0^{E_F} N(E)\,dE$, where E_F is the Fermi level, the highest occupied level). Figure 4.4b illustrates the distribution of occupied states in the conduction band (valence band) of a monovalent metal, such as sodium. The first band is only half filled and the distribution is approximately parabolic, as in the free-electron case. Figure 4.4c shows the distribution for a divalent metal, such as magnesium. As a result of overlapping of bands, electrons begin to fill the second, partially overlapping band before the first one is completely filled, giving two unfilled regions, both of which contribute to the electrical conductivity. The final sketch, Fig. 4.4d, describes the situation for a transition metal discussed earlier. In the free-electron model (Fig. 4.4a), $N(E)$ is proportional to $E^{1/2}$. The band models for monovalent and divalent metals show a similar behavior in their initial portions, but taper off to zero at the band edge. The cutoff on approaching the band edge is not sharp because in the three-dimensional case the values of k_x, k_y, and k_z, whose squares are proportional to E (Fig. 4.2), do not reach the forbidden region at the same value of E in all three directions in k space. The boundaries of the zones in k space are polyhedra and are called *Brillouin* zones. If the Brillouin zones were spherical, the cutoff for $N(E)$ would be sharp and $N(E)$ would be proportional to $E^{1/2}$ throughout the entire width of the band. Distribution curves for semiconductors and insulators are discussed in Section 5.1.

4.2 Chemisorption on Metal Alloys

The electron-band theory of metals has been applied mostly in attempts to explain the variation in chemisorption and in catalytic activity of alloys with composition, especially alloys of Group IB metals with transition metals. Changes in alloy composition alter the energy density of electrons at the Fermi level. In the particular case of transition metals, alloying with a Group IB metal also decreases the number of vacancies in the d band. Nickel–copper alloys have received much attention, and we shall use this system as an example of the filling of holes in the d band. In Section 4.1, we saw that the 3d and 4s bands of nickel metal overlap. These same bands also overlap in copper and to about the same extent. In nickel, the Fermi level lies slightly below the top of the 3d band (see Fig. 4.4d) and magnetism is due to the uncompensated spin moment of approximately 0.6 hole per atom in the d band. The magnetic moment of nickel is almost entirely due to the uncompensated spins of d electrons, and the intensity of magnetization in Bohr magnetons is approximately equal to the number of holes in the d band. The observed magneton number may not equal exactly the number of uncompensated spins because there is a small contribution from the orbital magnetic moment. The correction for nickel may reduce the number of holes in the d band to 0.55, but for our present purposes, we shall ignore the correction. Copper has one more electron than nickel and its d band is full. We assume that the single s electron of copper goes to the nickel d band, thus reducing the magnetic moment of nickel. Since the density of states in the d band of nickel is believed to be over ten times greater than that in the s band (see Section 4.1), the extra s electron per atom of copper should go at least 90% into the nickel d band. With 0.6 hole per atom in the nickel d band and a contribution of 1.0 electron per added atom of copper, the number of holes per atom of alloy as a function of copper content should be

$$
\begin{aligned}
n_{\mathrm{H}} &= (0.6 - x)/(1 + x), \qquad && x \leq 0.6, \\
n_{\mathrm{H}} &= 0, && x > 0.6, \qquad (7)
\end{aligned}
$$

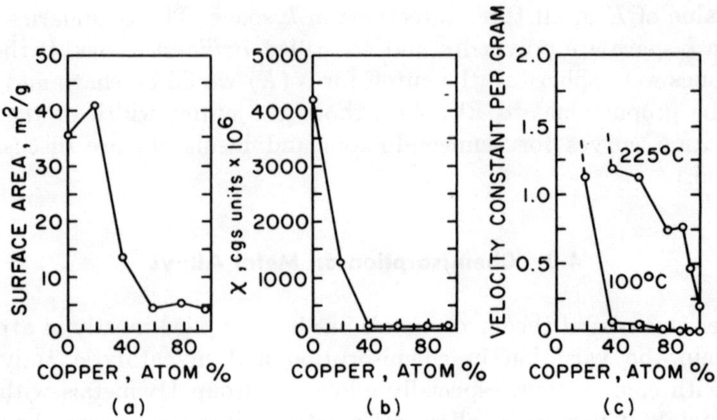

Fig. 4.5. Properties of skeleton powder nickel–copper catalysts. (a) Specific surface; (b) magnetic susceptibility; (c) activity for benzene hydrogenation. [After Reynolds [9]. By permission of The Chemical Society.]

where x is the number of copper atoms per nickel atom; or

$$n_{\mathrm{H}} = 0.6(1 - y) - y, \qquad y \leq 0.375,$$

$$n_{\mathrm{H}} = 0, \qquad y > 0.375, \tag{8}$$

where y is the atomic fraction of copper in the alloy and $(1 - y)$ is the atomic fraction of nickel. The magnetic moment of nickel would be expected to drop to zero at a copper–nickel ratio of 0.6 ($= x$), or at 0.375 atomic fraction of copper ($= y$).

How well do magnetic measurements on alloys confirm this simple picture? There are discrepancies in the literature about the experimental determination of the alloy composition at which the magnetic moment becomes zero. For example, Reynolds [9], who examined nickel–copper alloys prepared by exhaustive extraction of nickel–copper–aluminum alloy powders with sodium hydroxide, found that the magnetic moment had dropped to zero at about 40% copper in the alloy (Fig. 4.5b). When the same alloys were prepared by reduction of the oxides, however, he found that the magnetic moment dropped sharply to a low value at about 60% copper (Fig. 4.6b). Ahern *et al.* [10] used alloys produced by melting together pure nickel and copper in an induction furnace and reported an extrapolated value for zero magnetic moment at about 53% copper, compared to an extrapolated zero value at 60% reported by Best and Russell [11] (Fig. 4.7). Dowden and Reynolds [12] found zero magnetic

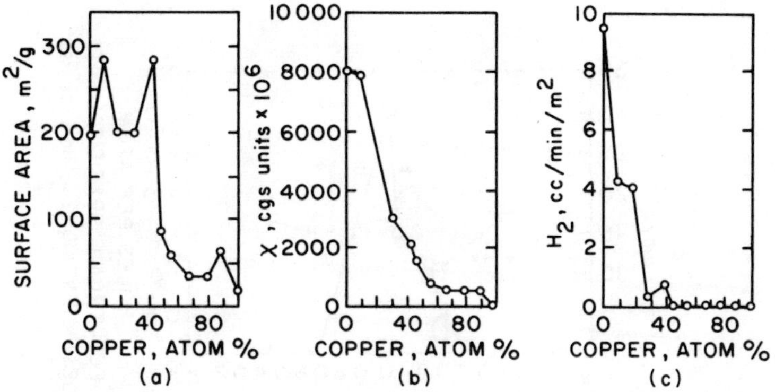

Fig. 4.6. Properties of nickel–copper catalysts prepared by reduction of the oxides. (a) Specific surface; (b) magnetic susceptibility; (c) specific activity for the hydrogenation of styrene. [After Reynolds [9]. By permission of The Chemical Society.]

moment at 60% copper (Fig. 4.8), in agreement with Best and Russell, and Reynolds [9] for reduced nickel–copper alloys (Fig. 4.6b). Both platinum and palladium are also believed to have about 0.6 hole in their d bands, and Couper and Eley [13] found saturation of the d band at about 60% gold in a palladium–gold alloy (Fig. 4.9). Couper and Metcalf [14] find the same saturation composition with palladium–silver alloys.

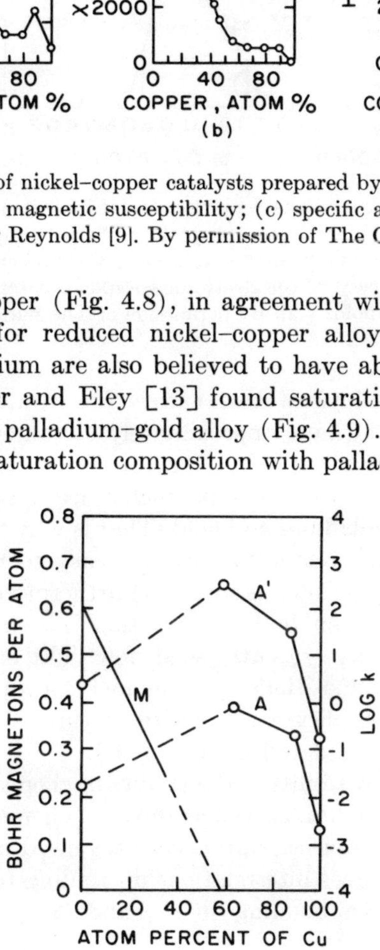

Fig. 4.7. Comparison of properties of the nickel–copper system as a function of catalyst composition. Curves A and A′, catalytic activity at −69°C and +3°C; curve M, saturation magnetic moment. [From Best and Russell [11]. By permission of The American Chemical Society.]

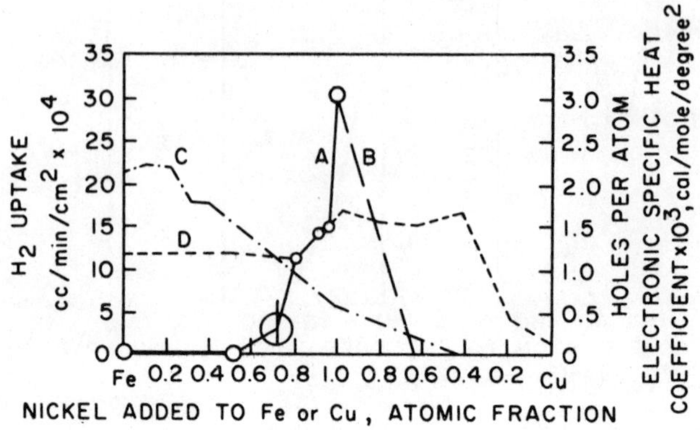

Fig. 4.8. Effect of nickel content on the rate of hydrogenation of styrene by alloy catalysts. Curve A, Ni–Fe catalysts, H_2 uptake $cm^3\ min^{-1}\ cm^{-2} \times 10^4$; curve B, Ni–Cu catalysts, H_2 uptake, $cm^3\ min^{-1}\ cm^{-2} \times 10^4$; curve C, number of holes per atom in the 3d band; curve D, coefficient of the electronic specific heat term $\times 10^3$, cal $mole^{-1}$ deg^2. [From Dowden and Reynolds [12]. By permission of The Faraday Society.]

In using binary alloys of this nature for catalytic studies, the two metals selected should have approximately the same lattice constants in order to eliminate this variable over the entire range of compositions. The lattice constants of copper and nickel are approximately the same, and so are those of palladium and gold. That is why these alloys have been selected. But homogeneous alloys are not easy to prepare, and in some instances the state of the alloy is suspect. Furthermore, surface composition may differ radically from bulk composition. For instance, Sachtler and co-workers [15–21] have recently shed new light on the relationships of bulk and surface compositions for the nickel–copper system, in which phase separations complicate the entire picture. We shall consider this important work in more detail in Section 4.4.

We consider now the results of the studies of chemisorption and catalysis on alloys. Catalytic studies, as well as those on chemisorption, are included because most of the empirical approaches reviewed in this section seem to assume a cause-and-effect interrelationship leading to a simple parallelism between chemisorptive bonding in d bands and subsequent catalytic reaction.

Using the skeletal nickel–copper alloys described above, Reynolds [9] found that magnetic susceptibility and catalytic activity for liquid–phase hydrogenation of benzene at 250 atm and 100°C fell continuously to zero over the range 0–40% copper, in agreement with simple d-band filling

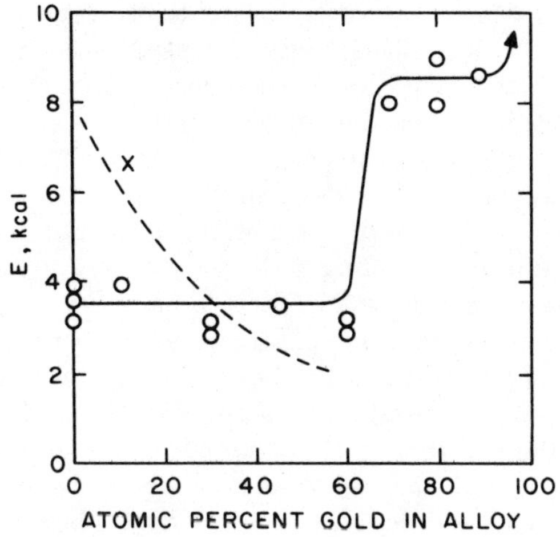

Fig. 4.9. The activation energy for the ortho–para hydrogen conversion on palladium–gold alloys as a function of composition (**13**). The broken line denotes the paramagnetic susceptibility in arbitrary units. [After Couper and Eley [13]. By permission of The Faraday Society.]

(Fig. 4.5). In the figure, it will be observed that catalytic activity fell more steeply than surface area. At a reaction temperature of 225°C, the results are not so clear-cut. Reduced oxide mixtures [9] showed similar declines in the hydrogenation of styrene, but did not reach low values of magnetic susceptibility until the composition had reached about 60% copper, while catalytic activity reached zero at 30–40% copper, similar to skeletal alloy (Fig. 4.6). Contrary to the work of Reynolds, Best and Russell [11] show a maximum in activity for the hydrogenation of ethylene at 60% copper over reduced nickel–copper catalysts prepared by the reduction of oxides formed from precipitated carbonates (Fig. 4.7). Pure nickel is known to chemisorb ethylene so strongly that the surface is poisoned. Coincident with the filling of d-band holes, a reduction in heat of adsorption of ethylene and hydrogen and an increase in catalytic activity may occur. Ultimately, as copper content is increased, d-band holes will become so sparse that activity will decrease. For the ortho–para hydrogen conversion at −196°C and at approximately room temperature, Shield and Russell [22] found little change in activity with alloy composition in the range from 5% to over 90% copper. But outside of this range, the changes were sharp, increasing with nickel content. Gharpurey and Emmett [23] investigated the hydrogenation of ethylene over nickel–copper alloy films and found a broad region from nearly pure nickel to about 80% copper where activity was relatively constant, while in the region of about 80% copper the activity increased markedly with increasing nickel content. These results are similar to those of Shield and Russell [22] for ortho–para hydrogen conversion, but do not agree with the results of Best and Russell [11] for hydrogenation of ethylene, as described above.

Dowden and Reynolds [12] have investigated the hydrogenation of styrene over nickel–copper and nickel–iron alloys. For the nickel–copper system, they found complete filling of d-band holes at an alloy composition of about 60% and a rapid decline in catalytic activity from pure nickel to an alloy composition of 40% copper (Fig. 4.8), which agrees with the results of Reynolds [9] (Fig. 4.6). They attribute the decline in activity to filling of the d-band holes. For nickel–iron alloys, both members of which have d-band vacancies, they introduce an additional factor to explain their results—the density gradient [$d \ln N(E)/dE$] at the Fermi level (see Section 4.1 for a description of $N(E)$). They consider that this factor may also contribute to the determination of the nature and the amount of adsorbed species; it is measured by the electronic component of the low-temperature specific heat. In the nickel–copper alloys, this factor remains nearly constant until all d-band holes have been filled (Fig. 4.8) and then declines precipitously. But in the nickel–iron system, it rises sharply in the region from 75 Ni–25 Fe to pure Ni, where the catalytic activity is also rising, even though the number of d-band holes is dropping continuously. Dowden and Reynolds associate the rise in activity with the rise in density gradient. We shall discuss this factor further in Section 4.3.

Couper and Eley [13] determined the catalytic activity of a series of palladium–gold alloys for ortho–para hydrogen conversion, using catalyst in the form of wires which are cleaned by bombardment with hydrogen atoms from an electric discharge and the dissolved hydrogen removed by heating to 400–600°C *in vacuo*. The activation energy remained low and approximately constant at about 3.5 kcal/mole from zero up to 60% gold (60 Au:40 Pd) and then rose sharply to about 8.5 kcal/mole. The magnetic susceptibility fell to zero at the point where the activation energy for ortho–para hydrogen conversion suddenly rose (Fig. 4.9), which, as we have seen, is not in agreement with the composition predicted by simple d-band filling. Activation energies for the same reaction on palladium–silver alloys [14] are qualitatively similar, but the change at 60% Ag is not so abrupt. In order to explain the nearly constant activation energy during the progressive filling of the d band, it was suggested that the necessary d-band vacancies are produced by excitation of electrons from d levels up to the Fermi level. Both alloy systems indicate that the incomplete d band of palladium-rich alloys is associated with low values of the activation energy and that the completed d band is associated with higher energies. But the association is not simple.

A study of the hydrogenation of 1,3-butadiene to butenes over palladium–gold alloys [24] revealed a sharp maximum in activation energy in the region 60–75% gold. Simple d-band filling is not fully adequate to explain the results, for activity continues after the d band, presumably, is filled. The authors suggest that in the composition range of d-band saturation, palladium atoms may be modified by strongly adsorbed butadiene so that they behave as if their d bands were partially filled. The same suggestion was made earlier in connection with the results of Best and Russell [11] for the hydrogenation of ethylene on nickel–copper alloys.

If no d-band vacancies exist in either constituent of a binary alloy, then the catalytic activity may depend entirely on the density gradient at the Fermi level. Schwab and co-workers [25–27] have used such alloys, chiefly silver alloyed with Zn, Cd, Hg, Ga, In, Tl, Sn, Pb, Sb, or Bi for the decomposition of formic acid into carbon dioxide and hydrogen. They reached the conclusion that chemisorption is favored by large, positive values of the density grandient $[d \ln N(E)/dE]$ at the Fermi level.

A general parallelism between d-band vacancies and chemisorption and catalysis does seem to exist, though there are exceptions. But the relationships are complex and never fully compatible with simply filling the d-band holes. Many of the discrepancies may be experimental; homogeneous alloys are not easy to prepare and surface compositions often do not reflect bulk compositions. Tightly adsorbed residues can act as poisons which plug d-band holes in transition metals just as nontransition metals do. In Section 4.4, we shall consider these and other critical points in detail.

4.3 The Theory of Chemisorption in d-Band Vacancies

We have mentioned before that the simplest theory merely states that chemisorptive or catalytic power drops as d-band vacancies drop. No quantitative relationship between the mutually downward paths, capable of verification, has been proposed. As we have seen, there are exceptions to the parallelism between chemisorption and d-band vacancies. There are even more exceptions to a monotonic relationship between chemisorption and catalysis. Yet the empirical nature of the original theory and its lack of detail make exceptions easy to explain. The ultimate goal is to develop quantitative theories of the chemisorptive state based on direct surface measurements—an impossible task, theoretically and experimentally, at present. Continued experiment and conjecture may lead to this goal, so long as conjecture is not congealed prematurely. It is difficult to conceive of chemisorption on transition metals that does not involve the d band, for in these metals the d band and the valence band (s band) overlap. The critical question is the nature of the involvement. In this section, we shall push the theory a little further, qualitatively.

Dowden [28] has considered the factors favoring positive ion, negative ion, and covalent bond formation on metal surfaces, and has given formal rules for determining bond characteristics. We summarize the rules. Positive ion formation is favored by (i) a large exit work function, (ii) a large positive value of the density gradient [$d \ln N(E)/dE$] at the Fermi level, $E_F = \phi$, and (iii) controlled quantities of electronegative additives which will raise E_F. A large work function obviously tends to attract electrons from the adsorbate into the d–s band and to hold them there. Also, a positive density of states gradient means that, as electrons added to the metal go to successively higher levels (raising the Fermi level), the density of electron states becomes greater. Negative ion formation is favored by (i) a low exit work function, (ii) a large negative value of the density gradient at the Fermi level, and (iii) controlled quantities of electropositive additives. A low exit work function tends to favor attraction of electrons to the adsorbate, and a negative value of the density of states gradient discourages the addition of more electrons by offering fewer states to them. Covalent bond formation, in which we are particularly interested here, is favored by (i) large values of the exit work function, (ii) large positive values of the density gradient, and (iii) the presence of d-band vacancies. These three rules for covalent bonding ensure favorable conditions for electrons from the adsorbate entering the metal band and for an increasing density of states to accommodate them. The bond may have some ionic character and may be either metallic (nonlocalized) or localized, depending on whether its energy lies inside or outside the upper and lower limits of the band energies of the metal.

A later publication by Dowden [29] describes chemisorption on metals in a manner analogous to the model of Friedel [30], who considered the system of a solute atom and its solvent. If the energy level (ionization potential) of an adsorbate electron, the s electron of hydrogen, for example, lies below the bottom edge of the metal band, then the proton is screened by a localized electron extracted from the bottom of the band. On the other hand, if the energy level of the s electron of the hydrogen atom lies above the bottom edge of the metal band, then the proton is screened by the mobile electrons in the band. Dowden assumes that the latter condition exists and gives the following relevant conclusions.

1. The donated s electrons enter the metal bands overlapping at E_F (Fermi level) in proportion to their respective densities, $N(E_F)$.
2. The radius R of the screening increases when the density of states at the Fermi level $N(E_F)$ decreases, when the density gradient becomes negative, and when the ratio of the number of donated electrons to the number of holes in the d band (both per metal atom) increases.

The first conclusion is succinctly stated; the second needs clarification. The screening radius R describes the effective range of the electric field resulting from the proton screened by electrons. It is usually expressed as the distance of a point from a charge screened by electrons where the value of the potential is a prescribed fraction of the unscreened value. When the density of the screening electrons is high, the screening radius R will be small, that is, the effective range of the field resulting from the proton charge is drastically reduced by the electrons. When the density of the screening electrons is low, R will be high. The unscreened proton will give the highest value of R. Thus, when the density of states $N(E_F)$ at the Fermi level is low, the density of screening electrons is low and the screening radius will be high. To understand why the screening radius increases when the density gradient becomes negative, we refer to Fig. 4.4d, the density of states diagram for nickel. We note that as electrons are added to the partially filled d band, the Fermi level rises to a higher energy value and that the value of $N(E)$ drops, that is, the density gradient is negative. With the decrease in $N(E)$, the density of the screening electrons is decreased, and consequently R increases. The conclusion that R increases with filling of the d-band vacancies is not really independent of the other two conclusions.

For copper, R is of the order of several atomic diameters because $N(E_F)$ is small. On the other hand, nickel has a large value of $N(E_F)$, as shown in Fig. 4.4d. But after filling of the d band by electrons from the adsorbate, the density of states $N(E_F)$ becomes low and consequently R is then large. If the nitrogen molecule uses only one electron per atom in bonding, which is apparently true at low temperatures [31], then the saturation coverage of nitrogen on transition metals should be close to that of hydrogen. But the dissociated nitrogen molecule gives adsorbed nitrogen atoms with three unpaired electrons, which are more than enough to fill the d-band vacancies per atom of nickel, cobalt, or iron. At saturation adsorption, R will be large and the coverage less than unity. The same general conclusions may be reached for tungsten, which has even more d-band vacancies.

Some nontransition metals can adsorb hydrogen, a fact that might indicate that d-band vacancies are a sufficient but not a necessary condition for chemisorption. Copper, for example, chemisorbs hydrogen at 300–400°C at a measurable rate with an activation energy of 20 kcal/mole [32]. The d-band theory is still invoked. Vacant d levels may be produced by raising a d electron to the top of the s band (see Fig. 4.4d), and the activation energy of chemisorption may be equal to the energy required for the d–s promotion. Similar explanations may hold for silver and gold, which have higher d–s promotional energies.

4.4 Status of Chemisorption in d-Band Vacancies

Two important questions are, What is the difference in chemisorption on transition and nontransition metals? Is the chemisorptive bond on transition metals localized or nonlocalized?

The first question calls for a knowledge of any differences in metal surface orbitals or bands between sp (nontransition) and d (transition) metals used in chemisorption. The chemisorptive bonds on sp metals are presumed to use metal s orbitals or sp^3 hybridized orbitals, in valence–bond language; or, in the language of band theory, s bands or overlapping sp bands. On d metals, the corresponding metal orbitals are presumed to be d orbitals or bands, or ds, dsp hybridized orbitals or overlapping bands. These orbitals or bands at the surface may be pictured as surface states distinct from crystal states, that is, surface bands lying outside any crystal bands or crystal orbitals considerably distorted. Alternatively, they may be pictured as "dangling bonds" or surface free valencies of atoms which at the surface do not have their normal coordination number. The precise meaning of a dangling bond is difficult to give. But in general, a dangling bond is described as a bond arising merely from the geometric presence of a surface, an extension into space of a metal orbital and not the result of new surface states. Adopting the latter view, we may say that the interaction of an s orbital on an adsorbate atom—hydrogen, for example—with an s orbital protruding from the metal surface forms what corresponds to a σ-molecular orbital in a diatomic molecule. In the case of interaction of an s orbital on the adsorbate with a p orbital on the metal surface, we assume that the interaction must be with the p_z orbital, the only one that protrudes from the surface, and the resultant bond is a σ bond. If there is hybridization of the s- and p-metal orbitals (corresponding to overlapping of s and p bands), the general character of the bonding orbitals is not altered. If the interacting adsorbate atom has p orbitals available, its p_z orbital may interact with the s and p_z orbitals extending from the metal surface to form σ-molecular orbitals. In addition, the p_x orbital of the adsorbate interacts with the p_x orbital of the metal, and similarly for the

p_y orbitals, to form localized π orbitals. These orbitals are doubly degenerate, and the situation is relevant to the interaction of oxygen on a metal surface, showing how double bonds may be formed. Interactions involving d orbitals of metal with adsorbate may be described in a similar manner.

Although this picture is probably a reasonable facsimile of the real picture of chemisorption on metals, it is not sufficiently quantitative to settle specific questions. It is not possible at present to obtain either wave functions or energies of the chemisorption states, and these values are critical to the differentiation between chemisorption on sp and d metals. The qualitative picture now available does not predict a drastic difference. Vacancies in the d band (or orbitals) do provide repositories for adsorbate electrons, but there are other ways of bonding. From molecular chemistry, we find, for example, that metals with filled d orbitals are among the best coordinating agents for olefins, indicating that back donation of electrons from filled d_{xz} and d_{yz} orbitals is at least as important as the filling of metal d_{z^2} orbitals by olefin π electrons.

If theoretical considerations do not provide a firm basis for distinguishing between chemisorption on sp and d metals, what does the experimental picture reveal? We summarize the situation:

1. All metals, except possibly gold, chemisorb oxygen;
2. sp metals chemisorb gases, such as CO and unsaturated hydrocarbons, at least as well as d metals, if not better;
3. d metals chemisorb hydrogen and nitrogen far more easily than sp metals.

It cannot be said from the data of Section 4.3 that a sharp division always occurs at the point where the d band is just full. The greater activity of d metals for the chemisorption of hydrogen does appear to indicate that d orbitals or bands are involved in some way. But the problem has become more complex with the finding [33–35] that preformed hydrogen atoms can be chemisorbed on Group IB metals, whereas molecular hydrogen cannot. This may indicate that chemisorption, in some cases, is limited by a large activation energy and not by the inability of the metal to chemisorb. The potential energy curves of chemisorption of hydrogen on sp and d metals may be quite similar. Dowden [36] points out that there is probably an initial state of chemisorption in many cases, called Type C chemisorption, which is intermediate between physical adsorption and strong chemisorption, and that there is a difference between the potential energy curves for this weak chemisorption. For transition metals, the Type C potential curve may intersect the curve for strong atomic ad-

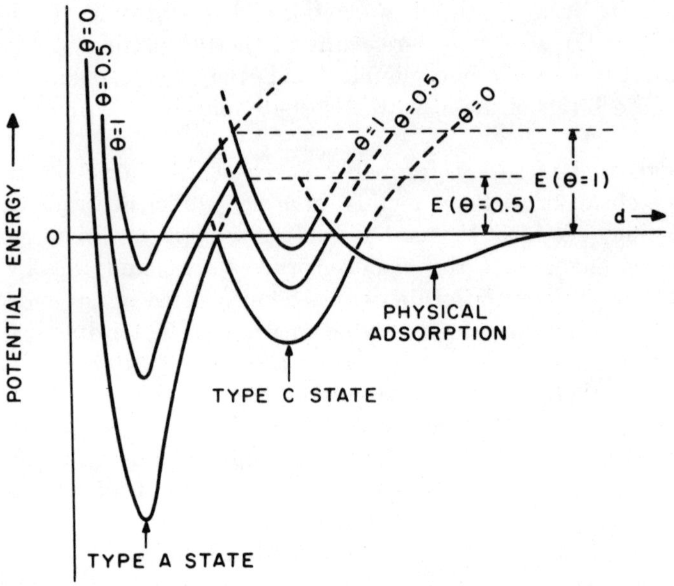

Fig. 4.10. Appearance of an activation energy between Type C and Type A states and the transition from the state of physical adsorption to Type C states. [After Gundry and Tompkins [41]. By permission of The Faraday Society.]

sorption in such a way that very little activation energy is required for the hydrogen to transfer to the final atom curve, Type A (Fig. 4.10). Evidence for weak chemisorption of this kind has been found [37–41]. In a nontransition metal, the crossing point may be located in a position that requires a large activation energy for the transfer.

The second question (whether chemisorption on transition metals is localized or nonlocalized), like the first question, has not been answered unequivocally yet. In an effort to find out what a quantum-mechanical treatment would predict concerning the nature of the bonding of hydrogen to a metal surface, Grimley [42] considered a simple model, a hypothetical metal with a nonoverlapping band system in which the Fermi electrons of the s band interact with the s orbital of a hydrogen atom. He used the LCAO–MO tight binding procedure (see Section 2.3), which includes a linear combination of metal s orbitals and the hydrogen 1s orbital. A calculation for the metal atoms alone showed the usual crystal band of energies and, under certain conditions (see Chapter 6), a band of surface states lying outside the range of crystal band energies. This surface band of energies decays exponentially into the metal perpendicular to the surface. When the hydrogen atom is brought into the system, Grimley shows that if surface states (bands) exist for the isolated metal, localized states will exist for the combined system and these states will decay exponentially perpendicular to the surface into the metal and also parallel to the surface. Two types of localized states are possible, one above the upper edge of the s band and one below the bottom edge. There will always be at least one localized state in the system. There may be a nonlocalized state with its energy in the range of s-band levels. To summarize, the possible combinations of localized states are

1. Two localized levels, one above and one below the s band.
2. Two localized levels, both below the s band.
3. Two localized levels, both above the s band.
4. One localized level below the s band.
5. One localized level above the s band.

The mathematical-physical rules governing which combination of states exists are complex. The important point is that there is always at least one localized state and there may be a nonlocalized state. For a given metal, both the number and type of localized states depend on the character of the interacting atom, and for a given atom they vary from metal to metal. The localized orbitals may spread over several atom diameters into the metal or along the surface about the central atom directly under the adsorbed atom. But the orbitals decay exponentially in all directions and the bond has the character of a σ bond in a diatomic molecule. If the localized orbital is doubly occupied, the bonding between the hydrogen atom and metal approximates closely a covalent bond.

How does this simple example relate to real metals—transition metals in particular? In real metals, bonds overlap and the energies of the orbitals of the localized system, hydrogen–metal, which might be considered as a surface compound, must lie outside all energy bands of the metal with which they interact. This is a severe restriction, particularly with band overlap, and could make localized adsorption impossible. Although the exact situation would seem to indicate that localized adsorption does not take place on real metals, we must remember that the calculation above shows that localization is only approximate, that there is a wave function extending throughout the crystal, but decaying exponentially. If certain d orbitals on a surface metal atom overlap orbitals of the adsorbate more strongly than they overlap the d orbitals on its neighboring atoms, the bonding electrons might become predominantly localized.

The conclusion from theoretical studies is that a situation approaching localized bonding in chemisorption on metals is possible. Only a detailed knowledge of the energy levels of the system and its components will

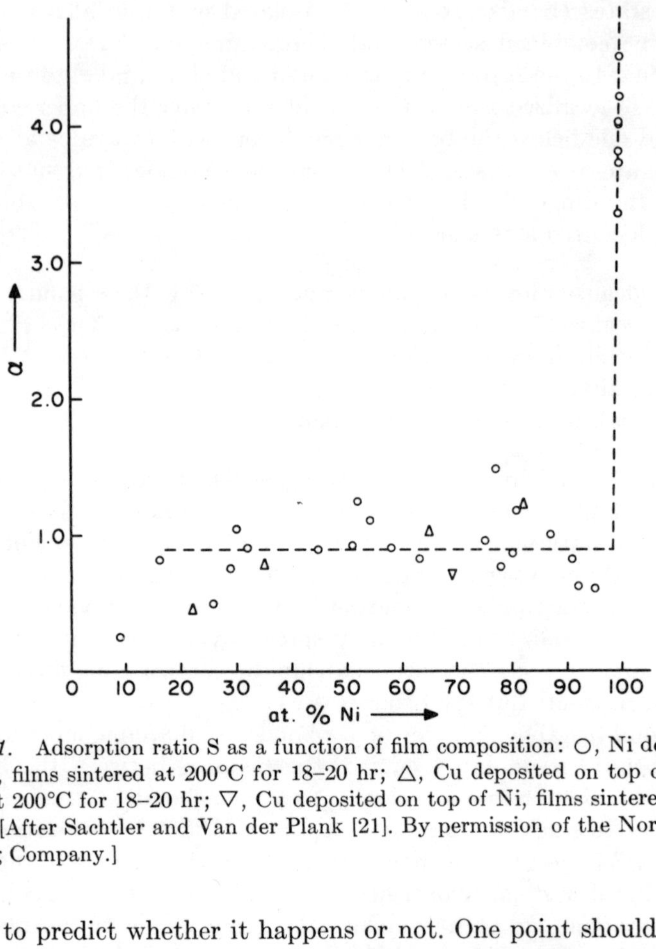

Fig. 4.11. Adsorption ratio S as a function of film composition: ○, Ni deposited on top of Cu, films sintered at 200°C for 18–20 hr; △, Cu deposited on top of Ni, films sintered at 200°C for 18–20 hr; ▽, Cu deposited on top of Ni, films sintered at 300°C for 14 hr. [After Sachtler and Van der Plank [21]. By permission of the North Holland Publishing Company.]

allow us to predict whether it happens or not. One point should be made clear: it is not always strictly meaningful to ask whether the collective properties of the band model or the localized properties of the individual surface atoms are decisive in forming nonlocalized or localized chemisorptive bonds, respectively. For we have seen that collective electrons may interact with adsorbate atoms to form either a localized bond or a nonlocalized one, depending on the quantum-mechanical properties of the system.

We turn to experimental work concerned with the problem of localized versus nonlocalized chemisorption on d metals. An important series of papers by Sachtler and co-workers [15–21] on this subject deals with chemisorption of hydrogen on copper–nickel alloys. They prepared copper–nickel films by successively depositing the two metals from the vapor and subsequently sintering the films under ultrahigh vacuum at 200°C. Photoelectric emission, X-ray diffraction studies, and data on diffusion led them to conclude that over a wide range of overall compositions (nearly pure Ni to approximately 80% Cu) the surface consists of a copper-rich alloy (80% Cu–20% Ni) and an almost pure nickel phase (2% Cu–98% Ni). In equilibrated films, each crystallite contains a kernel of nearly pure nickel enveloped in a skin of copper-rich alloy. Thus the composition of each phase is independent of the overall composition, and the surface properties of these alloys are constant, which is confirmed by experimental data on the work function. They measured the chemisorption of hydrogen at room temperature. In order to eliminate changes in chemisorption caused by changes in surface area, they expressed the amount of chemisorption as the ratio α of the number of hydrogen atoms chemisorbed to xenon atoms physisorbed at -196°C. The results are shown in Fig. 4.11. We note from the figure that:

1. The value of α for pure Ni is 3.9 ± 0.4. Assuming an area of 23 Å^2 per adsorbed xenon atom, each chemisorbed hydrogen atom occupies 5.9 ± 0.5 Å^2.
2. The value of α is approximately constant for alloys of constant surface composition.
3. Copper chemisorbs very little hydrogen at room temperature, considerably less than the alloys.
4. The ratio of the values of α for the alloys and for nickel is α alloy/α Ni $= 0.23 \pm 0.09$.

Sachtler and coworkers conclude that the first result agrees well with chemisorption of one hydrogen atom per surface nickel atom. The second result shows that the chemisorption of hydrogen is controlled by surface composition. The third result, they state, disproves that holes in the d band are decisive for chemisorption, since the copper-rich surface alloy (80% Cu–20% Ni) and pure copper are equally filled. It should be pointed out, however, that the energy levels or bands of surface atoms may differ in an unknown way from those of the bulk, so that there is some question about the number of d-band vacancies and their filling by copper. The fourth result indicates that hydrogen chemisorption is approximately proportional to the surface nickel concentration.

The results of Sachtler and co-workers support the localized model of chemisorption, just as do the theoretical studies of Grimley. The fourth result is not surprising, for in any valid model, localized or nonlocalized, an electron must be able to distinguish between copper and nickel. The third result may not quite disprove chemisorption in d-band vacancies.

It is conceivable that d–s promotion takes place, giving rise to chemisorption requiring an activation energy, which is similar to the explanation of the results for pure copper (see Section 4.3). Overall, their results are easier to understand if localized bonds are formed. It is perhaps too early to say that in general chemisorption on transition metals is not related to the presence of d-band vacancies [43–44]. Before that statement can be made, many more experimental and theoretical studies are needed, including data on the energies of surface states, which are hard to get, especially for metals.

REFERENCES

1. N. F. Mott and H. Jones, "The Theory of the Properties of Metals and Alloys." Oxford Univ. Press, London and New York, 1936.
2. A. H. Wilson, "Semiconductors and Metals." Cambridge Univ. Press, London and New York, 1939.
3. C. A. Coulson, "Valence," Chapter 11. Oxford Univ. Press, London and New York, 1953.
4. R. A. Levy, "Principles of Solid State Physics," Chapter 8. Academic Press, New York and London, 1968.
5. S. Raimes, "The Wave Mechanics of Electrons in Metals." North-Holland Publ., Amsterdam, American Elsevier, New York, 1970.
6. F. Bloch, *Z. Phys.* **52,** 555 (1928).
7. E. Wigner and F. Seitz, *Phys. Rev.* **43,** 804 (1933).
8. E. Wigner and F. Seitz, *Phys. Rev.* **46,** 509 (1934).
9. P. W. Reynolds, *J. Chem. Soc.* 265 (1950).
10. S. A. Ahern, M. J. C. Martin, and W. Sucksmith, *Proc. Roy. Soc. London* **A248,** 145 (1958).
11. R. J. Best and W. W. Russell, *J. Phys. Chem.* **76,** 838 (1954).
12. D. A. Dowden and P. W. Reynolds, *Discuss. Faraday Soc.* **8,** 184 (1950).
13. A. Couper and D. D. Eley, *Discuss. Faraday Soc.* **8,** 172 (1950).
14. A. Couper and A. Metcalfe, *J. Phys. Chem.* **70,** 1850 (1966).
15. P. Van der Plank and W. M. H. Sachtler, *J. Catal.* **7,** 300 (1967).
16. W. M. H. Sachtler, G. J. H. Dorgelo, and R. Jongepier, *J. Catal.* **4,** 100 (1965).
17. W. M. H. Sachtler and G. J. H. Dorgelo, *J. Catal.* **4,** 654 (1965).
18. W. M. H. Sachtler and R. Jongepier, *J. Catal.* **4,** 665 (1965).
19. W. M. H. Sachtler, G. J. H. Dorgelo, and R. Jongepier, *Proc. Int. Symp. Basic Probl. Thin Film Phys.* p. 218. Clausthal-Gottingen, 1965.
20. P. Van der Plank and W. M. H. Sachtler, *J. Catal.* **12,** 35 (1968).
21. W. M. H. Sachtler and P. Van der Plank, *Surface Sci.* **18,** 62 (1969).
22. L. S. Shield and W. W. Russell, *J. Phys. Chem.* **64,** 1592 (1960).
23. M. K. Gharpurey and P. H. Emmett, *J. Phys. Chem.* **65,** 1182 (1961).
24. B. J. Joice, J. J. Rooney, and P. B. Wells, and G. R. Wilson, *Discuss. Faraday Soc.* **41,** 223 (1966).
25. G. M. Schwab, *Trans. Faraday Soc.* **42,** 689 (1946).
26. G. M. Schwab and S. Pestamatjoglou, *J. Phys. Colloid Chem.* **52,** 1046 (1948).

27. G. M. Schwab, *Discuss. Faraday Soc.* **8,** 166 (1950).
28. D. A. Dowden, *J. Chem. Soc.* 242 (1950).
29. D. A. Dowden, *in* "Chemisorption" (W. E. Garner, ed.), p. 10. Butterworth, London and Washington D.C., 1958.
30. J. Friedel, *Advan. Phys.* **3,** 446 (1954).
31. O. Beeck, W. A. Cole, and A. Wheeler, *Discuss. Faraday Soc.* **8,** 314 (1950).
32. J. Kwan, *Res. Inst. Catal. Hokkaido* **1,** 95 (1949).
33. R. V. Culver, J. Pritchard, and F. C. Tompkins, *Z. Elektrochem* **63,** 141 (1959).
34. J. Pritchard and F. C. Tompkins, *Trans. Faraday Soc.* **56,** 540 (1960).
35. R. V. Culver, J. Pritchard, and F. C. Tompkins, *Proc. Int. Conf. Surface Activity* **2.** Butterworth, London and Washington D.C., 1957.
36. D. A. Dowden, "Chemisorption" (W. E. Garner, ed.), p. 5. Butterworth, London and Washington D.C., 1958.
37. A. Eucken and W. Hunsmann, *Z. Phys. Chem.* **B44,** 163 (1939).
38. F. C. Tompkins, *Z. Elektrochem.* **56,** 360 (1952).
39. R. A. Beebe and N. P. Stevens, *J. Amer. Chem. Soc.* **62,** 2134 (1940).
40. R. A. Beebe and D. A. Dowden, *J. Amer. Chem. Soc.* **60,** 2912 (1938).
41. P. M. Gundry and F. C. Tompkins, *Trans. Faraday Soc.* **52,** 1609 (1956).
42. T. B. Grimley, "Chemisorption" (W. E. Garner, ed.), p. 17. Butterworth, London and Washington D.C., 1958.
43. Z. Bastl, *Surface Sci.* **22,** 465 (1970).
44. W. M. H. Sachtler, *Surface Sci.* **22,** 468 (1970).

V

Electron-Band Theory and the Ionic Bond

In this chapter, we shall consider ionic chemisorption on semiconductors, characterized by electron transfer between the adsorbate and the conduction band or valence band of the solid. The theories discussed here are a step beyond the analogies with simple molecular bonds of Chapter III and a step toward a complete quantum-mechanical approach. Foreign atoms are brought into interaction with the band structure of a semiconductor, just as was done for metals in Chapter IV. In later chapters, the wave functions of the foreign atom and the atoms of the solid are merged at the outset into an integrated system. The chapter opens with a qualitative description of the electron theory of semiconductors and insulators. Semiconductors are broken down into intrinsic and extrinsic types. More important in chemisorption and catalysis are the extrinsic semiconductors, which rely on energy levels created by imperfections or impurities in the lattice. Extrinsic semiconductors are broken down further into n type and p type, which conduct through electrons in the conduction band or holes in the valence band, respectively. In both types, the significance of the Fermi level, the distribution of energy levels, and the various kinds of imperfections are described. In the second section, chemisorption on these structures is considered. We shall describe the *boundary-layer* theory, which follows closely the theory of the contact between a metal and a semiconductor, and applies strictly to ionic adsorption. As a result of the charge on adsorbed particles, a space charge develops in the semiconductor and there exists at the surface of the semiconductor a boundary layer across which there is an increasing or decreasing electron concentration in going from the inner edge of the layer toward the surface. A few more sophisticated approaches of a statistical or quantum-mechanical nature, still within the guidelines of this chapter, will be outlined. In the final section, experimental studies of ionic complexes and the changes in semiconductivity with ionic adsorption will be reviewed.

5.1 Semiconductors and Insulators†

The band structure of metals belongs to one of the following categories: an isolated, partially filled valence band; a partially filled valence band overlapping an empty band; a full valence band overlapping an empty band. On the other hand, all bands in insulators and semiconductors are either full or empty and isolated. The highest full band is called the valence band and the empty band above it, separated by an energy gap, is called the conduction band. In the case of metals, conduction occurs in the valence band, for the valence band is either partially filled or it overlaps an empty band. An insulator and an intrinsic semiconductor are dis-

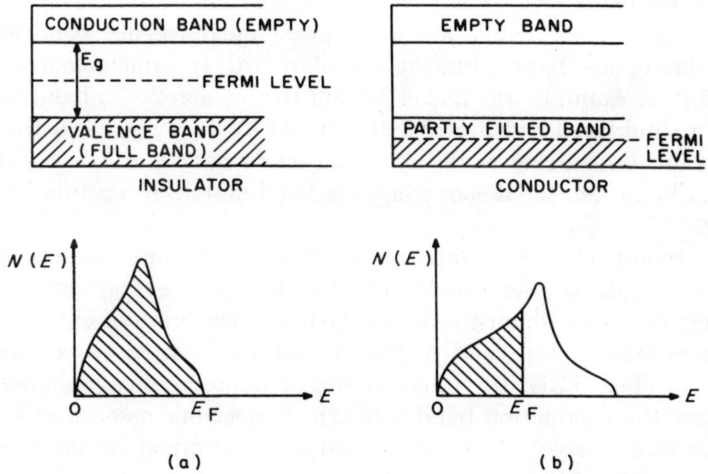

Fig. 5.1. Insulator or semiconductor. Filled band in (a) makes it an insulator when E_g is large and an intrinsic semiconductor when E_g is small. Partially filled band in (b) makes it a conductor. Bottom figures give the corresponding density of states curves.

† Refs. [1–5].

tinguished by the breadth of the energy gap which separates the valence band and the conduction band, as shown in Fig. 5.1. When the band gap E_g is large, as in CaO, which has a band gap of about 10 eV [6], the substance is an insulator. For intrinsic semiconductors, the band gap is much smaller, usually below 1 eV. If the energy gap is small, a reasonable number of electrons can be excited from the valence band across the energy gap to the conduction band, where conduction will occur because the band is sparsely filled. Conduction also occurs by positive holes left in the valence band. No conduction occurs in a filled valence band, since the flow of electrons in one direction is balanced by an equal flow in the other, resulting in no net flow.

We shall consider chiefly compound semiconductors, especially metal oxides, because they are most significant in chemisorption and catalysis. For such oxides, there is a question which element forms the band structure of interest. In heteropolar semiconductors, the metal ions will usually form the deeper energy states and there will be very little overlapping of the wave functions of neighboring metal ions. Thus, broadening of the energy states of the metal ions into bands is usually insignificant. But oxygen bands are formed more easily because oxygen ions and anions in general are larger than metal ions. The overlapping filled 2s and 2p levels of isolated ions (O^{2-}) are filled and separated from overlapping 3s and 3p states. In some cases, significant broadening of metal ion bands may occur and such bands or levels may overlap oxygen bands. Whatever the situation, the important bands are the highest filled valence band and the empty conduction band immediately above it. In stoichiometric nickel oxide [6], for example, the nickel 3d and the oxygen 2s–2p bands overlap to give a total band width of 0.2 Ry (13.54 eV = 1 Ry), and the nickel 3d band alone in nickel oxide is calculated to have a width of 0.13 Ry. For metallic nickel, the d–s overlapping band system has a total width of 0.37 Ry.

Far more important than intrinsic semiconductors for chemisorption and catalysis are extrinsic semiconductors, which rely on energy levels created by imperfections or impurities in the lattice. Two types of extrinsic semiconductors exist: *n* type and p type. In n-type (normal or excess) semiconductors (Fig. 5.2a), the imperfection or impurity creates donor levels just below the conduction band into which electrons associated with the impurity atom may be excited. In p-type (abnormal or deficit) semiconductors (Fig. 5.2b), acceptor levels lie just above the valence band. Electrons are excited from the valence band into the acceptor impurity levels, and conduction occurs as a result of the positive holes left in the valence band. Since the impurity levels are small in number and widely

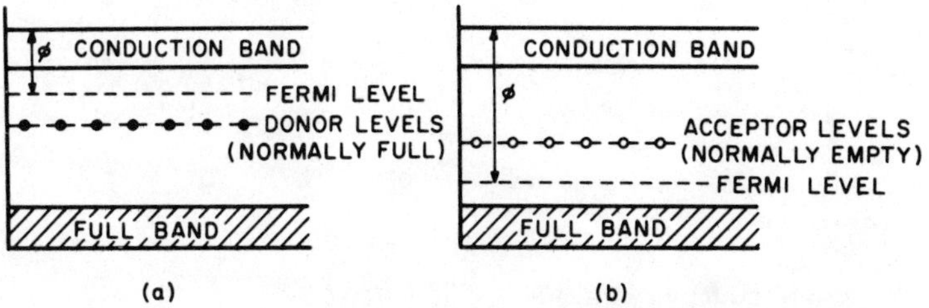

Fig. 5.2. Types of semiconductors: (a) n-type; (b) p-type.

separated in the crystal, they are essentially discrete, so that conduction does not occur in them. The energy gap between the donor level and the conduction band or the valence band and the acceptor level is much smaller than the gap between the valence band and the conduction band. For example, the band gap in ZnO, an n-type semiconductor, is 3.2 eV, but the impurity gap (donor level to conduction band) is only about 0.02 eV; and in TiO_2 (also n type), the corresponding values are 3.0 and 0.07 eV [4]. Values for p-type semiconductors are in the same range.

Many types of imperfections affect conduction in crystals. There are reversible imperfections, consisting of Frenkel defects, in which interstitial atoms occur, and Schottky defects, in which vacant lattice sites occur. There are irreversible imperfections, such as Smekal cracks and dislocations. But the imperfections of most significance for chemisorption and catalysis are those irreversible imperfections which arise from chemical rather than physical origins. They stem from nonstoichiometric compositions or from the presence of foreign ion impurities. We shall give a simplified review of them. They may be classified as follows:

1. anion vacancies;
2. interstitial cations;
3. cation vacancies;
4. interstitial anions;
5. addition of higher-valent foreign ions to n- and p-type semiconductors;
6. addition of lower-valent foreign ions to n- and p-type semiconductors.

Heteropolar crystals with anion vacancies or interstitial cations show a deficiency of anions, and heteropolar crystals with cation vacancies or interstitial anions show a deficiency of cations. Because of the large size of anions, interstitial anions are seldom, if ever, observed. Current in crystals with anion vacancies and interstitial cations is carried by electrons. The electrons associated with these defects are not as tightly bound as they would be to oxygen in stoichiometric crystals, and therefore they

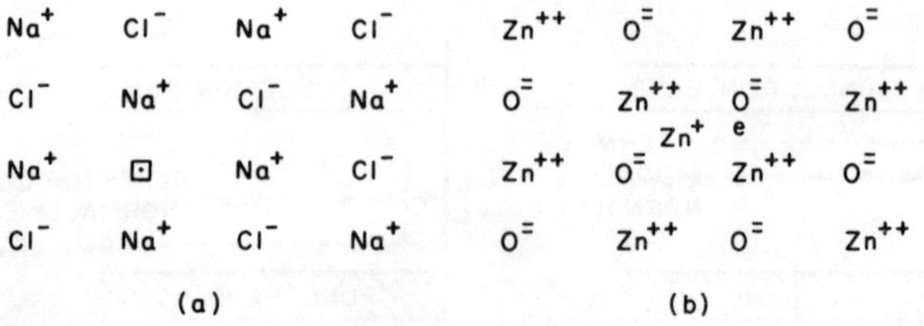

Fig. 5.3. Semiconductors with chemical imperfections. (a) NaCl n-type with electron trapped in anion vacancy, ⊡, occupying donor impurity level; (b) ZnO n-type with interstitial Zn^+ and quasi-free electron e occupying donor impurity level.

occupy a higher energy level (<0). In fact, their level is very close to the bottom of the conduction band, as is typical of n-type semiconductors (Fig. 5.3a,b). As the temperature is raised, electrons jump into the conduction band, where they are essentially free and conducting. They cannot conduct in the impurity level because these are discrete levels. There is one level for each defect ion, and all the discrete levels are degenerate, that is, have the same energy. They are discrete because defect ions are few in number and thus too widely separated to form bands of energy. When an electron is trapped in an anion vacancy (the impurity level), the site is called an *F* center; and when a second electron is trapped in the same vacancy, the site is called an *F′* center.

Current in crystals with cation vacancies is carried by positive holes (Fig. 5.4). Such crystals possess excess anions and a dearth of cations. In order to preserve electrical neutrality of the crystal as a whole, electrons are taken from the cations and given to the excess of negative constituent. In the figure, two of the copper ions are shown as divalent, since each has supplied an electron to satisfy the extra oxygen. The extra positive holes, that is, empty levels in the otherwise full levels, are discrete, just like impurity levels in n-type semiconductors, and lie in this case just above the valence band. There is a net movement of positive holes through the valence band, opposite in direction to the movement of electrons, which in calculations may be treated like the movement of electrons but opposite

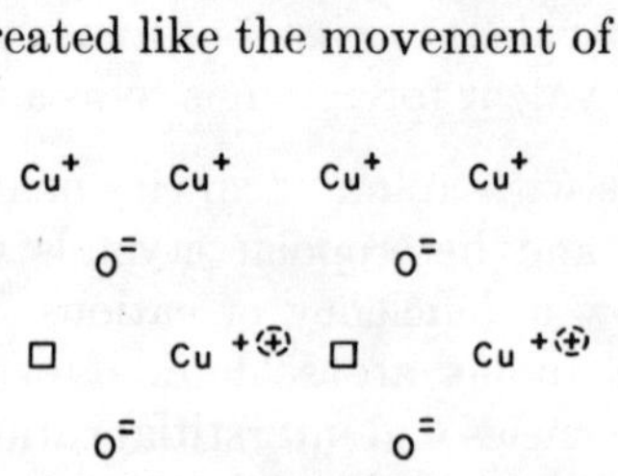

Fig. 5.4. Semiconductor with chemical imperfections, Cu_2O p type with cation vacancies and quasi-free positive hole ⊕ at acceptor impurity level.

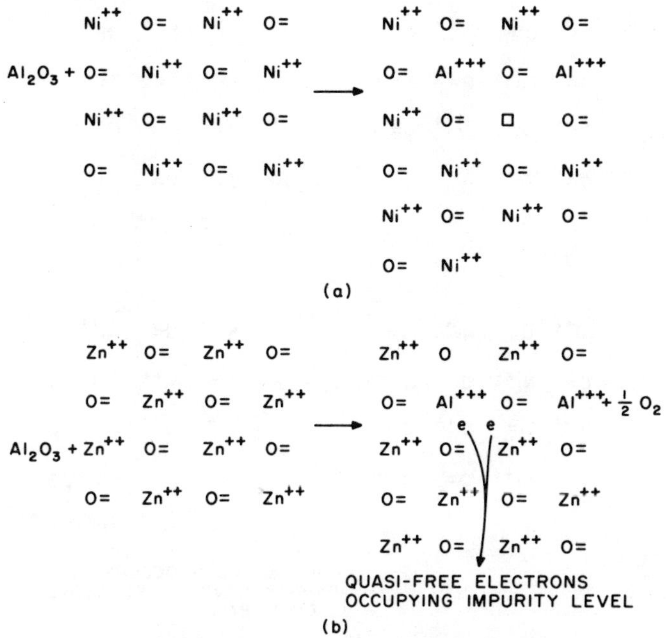

Fig. 5.5. Semiconductors with chemical imperfections. (a) NiO p type with higher-valent foreign ion and cation vacancy, nonconducting; (b) ZnO n type with higher-valent foreign ion and quasi-free electrons just below the conduction band, conducting.

in sign. A semiconducting crystal in which current is carried by positive holes, as we have mentioned, is p-type.

Similarly, the introduction of foreign ions into a crystal can produce or enhance n- or p-type conductivity. Adding a higher-valent cation to a crystal lattice produces either cation vacancies or quasi-free electrons. When the higher-valent cation, Al_2O_3 for example, is added to a lower-valent p-type semiconductor, such as NiO, cation vacancies with no positive holes are probably formed (Fig. 5.5a). This form is nonconducting. On the other hand, when Al_2O_3 is added to a lower-valent n-type semiconductor ZnO, quasi-free electrons are produced which give conduction when excited from donor levels into the conduction band. Whether cation vacancies or quasi-free electrons are formed depends on the complex energy relationships within the crystal.

Adding a lower-valent cation to a crystal produces either anion vacancies or positive holes. The addition of Li_2O to ZnO, for example, produces an anion vacancy with no electrons (Fig. 5.6a), and the crystal is nonconducting, or at least exhibits no increase in conductivity over ZnO

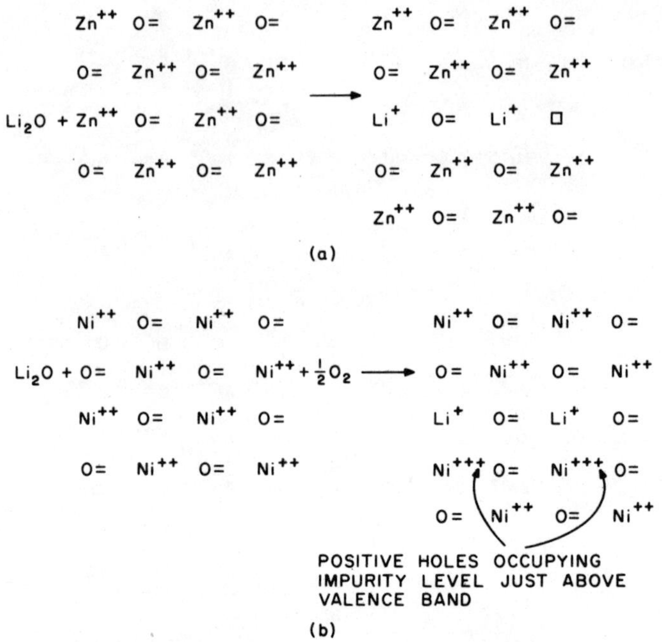

Fig. 5.6. Semiconductors with chemical impurities. (a) ZnO n type with lower-valent foreign ion and an anion vacancy, nonconducting; (b) NiO p type with lower-valent foreign ion and positive holes just above the valence band, conducting.

alone. But the addition of Li_2O to NiO increases conductivity, producing positive holes just above the valence band.

We turn to a discussion of the Fermi level in semiconductors. An intrinsic semiconductor normally contains very few electrons in the conduction band because only those at the top of the valence or filled band can receive enough energy to bridge the large gap. Only the lower levels of the conduction band will be occupied, so that electrons in these levels may be considered to behave as free electrons, which means that Fermi–Dirac statistics are applicable. From the energy density of states (see Section 4.1) and the probability f that a state will be filled, the number of electrons in the conduction band, n_e, is calculated. Similarly, the number of holes in the valence band, n_h, is calculated. (We use the fact that $1 - f$ is the probability that a state is empty, i.e., has positive holes.) The two expressions may be equated, for $n_e = n_h$, and solved for the Fermi level E_F. Within the free-electron approximation, the Fermi level lies in the energy gap, midway between the top of the valence band and the bottom of the conduction band (see Fig. 5.1a). More exact considerations show that the

Fermi level will lie indeed between the two bands, but not necessarily at the midpoint. Similarly, the Fermi level for defect semiconductors will lie approximately midway between the discrete impurity level (donor or acceptor) and the bottom of the conduction band (n type) or the top of the valence band (p type) (see Fig. 5.2).

We conclude this section with a brief discussion of the theory of conductivity in semiconductors, which is important in the study of chemisorption. Conductivity is expressed in general by

$$\sigma = nev \tag{1}$$

where n is the number of conducting electrons per unit volume or holes in the valence band, v is their mobility, and e their charge. If v is constant, as in a given substance, conductivity becomes proportional to the number of conducting electrons n. Using Fermi–Dirac statistics, the number of conducting electrons in an intrinsic semiconductor is found to be

$$n = (4\pi mkT/h^2)^{3/2} \exp(-E_g/2kT) \tag{2}$$

where E_g is the width of the energy gap from the top of the filled band to the bottom of the conduction band. We may write for conductivity

$$\sigma = A \exp(-E_g/2kT) \tag{3}$$

where A is a constant for a given substance. The linearity of a plot of $\log \sigma$ versus $1/T$ has been confirmed in many cases for both intrinsic and extrinsic semiconductors. However, the plot for extrinsic semiconductors often shows a broken line that can be expressed as the sum of two straight lines on a $\log \sigma$ versus $1/T$ plot,

$$\sigma = A_1 \exp(-E_1/kT) + A_2 \exp(-E_2/kT), \tag{4}$$

where one member on the right-hand side is presumed to represent the intrinsic semiconductivity of electrons excited from the filled band to the conduction band and the other defect conductivity of electrons excited from impurity levels to the conduction band. The activation energy for defect conductivity is much smaller than that for intrinsic conductivity.

5.2 Theories of Chemisorption on Semiconductors

The Boundary-Layer Theory

The boundary-layer theory of chemisorption was developed in the early 1950s chiefly by Aigran and Dugas [7], Hauffe and Engell [8, 9], and Weisz [10, 11], and for many years held the spotlight in studies of chemi-

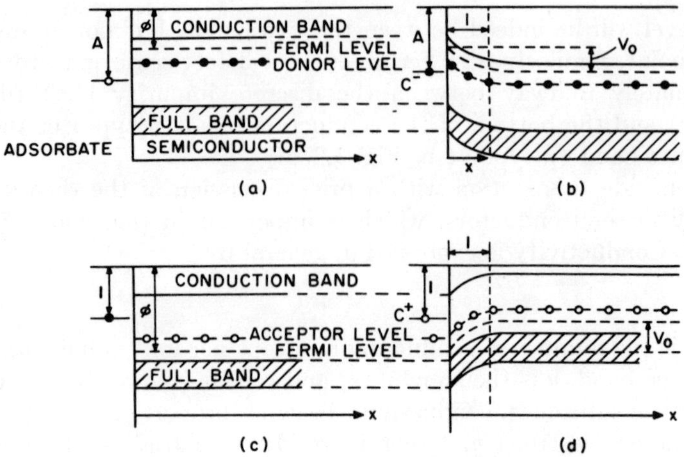

Fig. 5.7. Adsorption of C as C^- on n-type semiconductor (a) before adsorption; (b) after adsorption. Adsorption of C as C^+ on p-type semiconductor (c) before adsorption; (d) after adsorption.

sorption and catalysis. It follows closely the earlier theory of the contact between a metal and a semiconductor, a good description of which is given by Mott and Gurney [1]. We consider four cases: chemisorption as negative and as positive ions on n- and p-type semiconductors. We consider first an atom C which is adsorbed as a negative ion C^- on an n-type semiconductor. A good example is the adsorption of oxygen as either O^- or O^{2-} on zinc oxide,

$$\tfrac{1}{2}O_2 + 2e \rightarrow O^{2-} \tag{5}$$

In Fig. 5.7a,b, the situation is pictured before and after adsorption of C. Before adsorption, the electron-accepting level of C is shown in the region below the bottom of the conduction band and above the top of the valence or full band. When an atom C is brought to the surface of the semiconductor, an electron from the conduction band is transferred to the electron-accepting level of C, creating the negative ion C^-. Gradually, as adsorption proceeds, a positive space charge develops in the semiconductor as a result of the loss of electrons to atoms of C, and a negative charge builds up on the surface. This is the boundary layer. A field exists in this layer that forces electrons away from the surface into the interior. The force dies off with distance, and a short distance into the metal, the distribution of electrons in the conduction band and discrete impurity levels has returned to normal. Thus, the boundary layer is a region of decreasing electron concentration in going from the inner edge of the layer toward the surface of the semiconductor. As the space charge accumulates with continued adsorption, it becomes more and more difficult to transfer electrons from the conduction band to C. When the meager supply of electrons in the conduction band is used up, a further supply would have to come from the much deeper levels of the valence band, which may require a prohibitively large activation energy. Chemisorption could cease far short of a monolayer as the supply of electrons decreases and the space charge builds up. To this type of adsorption, Hauffe has given the name *depletive chemisorption*. The work function ϕ, which is the energy required to remove an electron from the Fermi level of the semiconductor to infinity, increases by the amount of additional energy required to move the electron across the potential V of the boundary layer. Effectively, the Fermi level is pushed down as adsorption proceeds. When the potential energy of the electrons, as represented by the Fermi level, becomes equal to the potential energy of the electron-accepting level of C (electron affinity A_0 of C), equilibrium is established. With loss of electrons from the conduction band, conductivity decreases. Both limited adsorption and fall in conductivity have been observed [12].

We look now at the similar case of adsorption of a negative ion C^- on a p-type semiconductor, which is exemplified by the adsorption of oxygen as O^- on nickel oxide. In this case, electrons must come from the valence band, and the top of this band must lie considerably higher than it does in the case of n-type semiconductors, such as zinc oxide. Otherwise, the activation energy for chemisorption would be too high. Given a reasonable activation energy, there is an ample supply of electrons near the top of the valence band, for this band, unlike the conduction band, is initially full. As electrons are removed from the valence band, the positive holes left in the band will cause increased conductivity. Adsorption of this type is called *cumulative chemisorption*, for it is not limited as depletive chemisorption is. High adsorption accompanied by increased conductivity has been observed [13, 14]. The band picture is similar to that in Fig. 5.7a,b for C^- on an n-type semiconductor except that the large supply of electrons near the surface means that the potential barrier to further adsorption is lower and the width of the barrier less. The Fermi level is effectively raised in this case.

When C loses an electron, becoming C^+, a corresponding picture on n- and p-type semiconductors may be developed. On an n-type semiconductor, an electron removed from C (ionization potential I) enters the conduction band (electron affinity ϕ), increasing the conductivity. Since the conduction band is nearly empty, large numbers of electrons can be put into it and monolayer coverage should be attained, in accordance with

TABLE 5.1

Types of Chemisorption According to the Boundary-Layer Theory

Adsorbed species	Semiconductor	Types of adsorption
C^-	n type	Depletive
C^-	p type	Cumulative
C^+	n type	Cumulative
C^+	p type	Depletive

cumulative-type chemisorption. On the other hand, chemisorption on a p-type semiconductor belongs to the depletive type, for electrons from C enter the nearly full valence band. When holes in the valence band levels near the surface of the crystal are filled, further adsorption takes place only by crossing the increasing potential barrier to valence band levels deeper in the crystal. In this case, the Fermi level is pushed up (Fig. 5.7c,d) and the valence and conduction bands curve downward on approaching the surface across the potential barrier. In Table 5.1, we summarize the types of ionic chemisorption systems according to the boundary-layer theory.

The quantitative development of the boundary-layer theory follows from the theory of the contact between a metal and a semiconductor. Assuming equilibrium conditions, we use the first two types in Table 5.1 as models. Let x be the maximum distance of any point in the boundary layer from the surface of the semiconductor, and $x = l$ the maximum width of the layer. If $V(x)$ is the potential difference between the interior of the semiconductor and point x, then $V(0) = V_0$ is the potential difference at the surface and $V_l = 0$. We want to find expressions for $n_e(x)$ and $n_h(x)$, the densities of free electrons in an n-type semiconductor and positive holes in a p-type semiconductor, respectively, at point x. When x is greater than l, n_e and n_h will have the constant values $n_e(l)$ and $n_h(l)$ characteristic of the semiconductor. Assuming one electron per defect, the maximum values of the densities $n_e(l)$ and $n_h(l)$ are designated n_d and are attained when all donor levels have lost their electrons to the conduction band or all acceptor levels have received an electron each from the valence band. If the density of electrons or holes is so low that it can be assumed electrons are all in the lowest level for practical purposes and that Pauli's principle of accommodation of electrons and the use of Fermi–Dirac statistics can be bypassed, then it is permissible as an approximation to use Boltzmann statistics (see Section 3.1). We find

$$n_e(x) = n_e(l) \exp(-eV(x)/kT), \tag{6}$$

$$n_h(x) = n_h(l) \exp(eV(x)/kT). \tag{7}$$

When $n_e(x)$ and $n_h(x)$ are in the boundary layer $(0 < x < l)$, they will behave as shown in Figs. 5.8a and c, respectively. But Hauffe and Engell [8, 9] simplified the problem by assuming that $n_e(x)$ and $n_h(x)$ are constant in the boundary layer, as shown in Figs. 5.8b and d. We see from Eqs. (6) and (7) that this approximation applies best when eV_0 is very much greater than kT, for then $n_e(0)$ is very small compared to $n_e(l)$ and $n_h(0)$ very large compared to $n_h(l)$. The space charge ρ (dearth of free electrons in the conduction band or excess holes in the valence band compared to the interior of the crystal) in the boundary layer will also be constant with these assumptions and is expressed by

$$\rho = e[n_e(l) - n_e(0)] \cong en_e(l), \qquad 0 < x < l, \tag{8}$$

$$\rho = e[n_h(0) - n_h(l)] \cong en_h(0), \qquad 0 < x < l, \tag{9}$$

$$\rho = 0, \qquad x \geqslant l. \tag{10}$$

If every adsorbed atom captures one electron and the concentration of

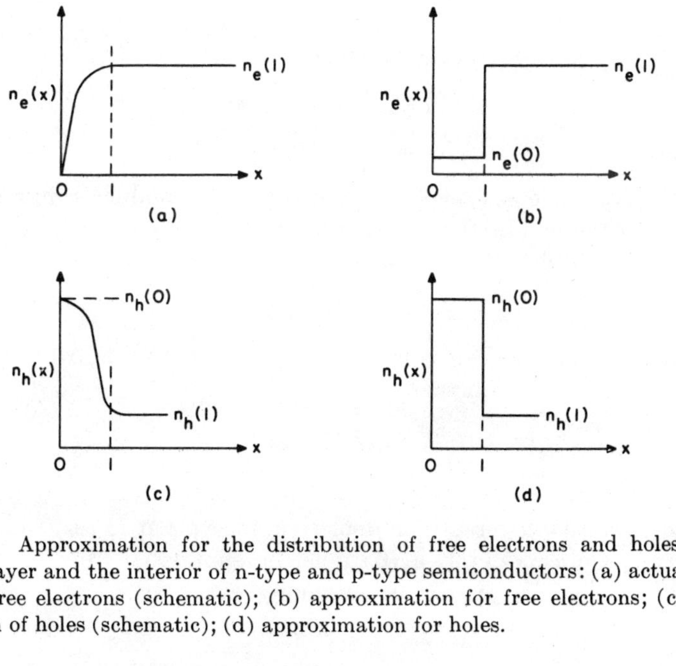

Fig. 5.8. Approximation for the distribution of free electrons and holes in the boundary layer and the interior of n-type and p-type semiconductors: (a) actual distribution of free electrons (schematic); (b) approximation for free electrons; (c) actual distribution of holes (schematic); (d) approximation for holes.

un-ionized defects is the same in the interior and in the boundary layer, then the number of surface charges per unit area, that is, the number of chemisorbed atoms, is

$$N_s = [n_e(l) - n_e(0)]l = ln_e(l), \tag{11}$$

the number of electrons in a boundary-layer volume $1 \times 1 \times l$. And for a p-type semiconductor

$$N_s = [n_h(0) - n_h(l)]l = ln_h(0). \tag{12}$$

We know from the Poisson equation, which holds in a space-charge layer whether equilibrium has been established or not, that

$$d^2V(x)/dx^2 = 4\pi\rho/\epsilon, \tag{13}$$

where ϵ is the dielectric constant. Integrating with the boundary conditions $x = l$, $dV/dx = 0$, and $V = 0$, and setting $x = 0$ in the result, we find

$$V_0 = 2\pi\rho l^2/\epsilon. \tag{14}$$

Solving for l in Eqs. (11) and (12) and substituting in Eq. (14), we have

$$V_0 = 2\pi\rho N_s^2/\epsilon n_e^2(l), \tag{15}$$

$$V_0 = 2\pi\rho N_s^2/\epsilon n_h^2(0). \tag{16}$$

Finally, substituting for ρ from Eqs. (8) and (9), and solving for N_s, we obtain

$$N_s = [V_0\epsilon n_e(l)/2\pi e]^{1/2}, \tag{17}$$

$$N_s = [V_0\epsilon n_h(0)/2\pi e]^{1/2}. \tag{18}$$

If every donor defect level in an n-type semiconductor has yielded its electron to the conduction band, then in Eq. (17) $n_e(l)$ becomes n_d, the total number of defects per unit volume, and we write

$$N_s - [V_0\epsilon n_d/2\pi e]^{1/2}. \tag{19}$$

Assuming reasonable values for V_0 of $1V$; ϵ, 10; and n_d, $10^{18}/\text{cm}^3$, we find for depletive adsorption that

$$N_s = 2.5 \times 10^{12}/\text{cm}^2. \tag{20}$$

The number of atoms in the surface of a metal oxide is of the order of 10^{15}, so that the coverage for depletive adsorption is less than 1%. For the p-type semiconductor, $n_h(0)$ must be eliminated from Eq. (18) with the aid of Eq. (7) at $x = 0$. Assuming every acceptor level has been filled, then $n_h(l)$ goes to n_d, and we obtain

$$N_s = [V_0 \epsilon n_d/2\pi e) \exp(eV_0/kT)]^{1/2}. \tag{21}$$

Note the extra exponential term in this equation compared to Eq. (19) for depletive chemisorption on an n-type semiconductor. The extra term permits greater coverage, as we would expect for the cumulative chemisorption of C^- on a p-type semiconductor.

It is interesting to note that the boundary-layer mechanism of chemisorption inherently prescribes a decrease in energy of the chemisorptive bond with coverage. When C^- forms on the surface of an n-type semiconductor, the initial energy of adsorption is $A_0 - \phi + \beta$, where A_0 is the electron affinity of C, ϕ is the work function of the solid, and β is the interaction energy (electrostatic attraction, interaction integral, etc.). As adsorption proceeds, the potential barrier V_0 builds up and the work function increases by this amount, until at equilibrium we have $A_0 - (\phi + V_0) + \beta = 0$, and further adsorption is precluded. A similar argument holds when C^+ forms on a p-type semiconductor. The initial energy in this case is $\phi - I + \beta$, where I is the ionization potential of C. Chemisorption continues until $\phi - (I + V_0) + \beta = 0$.

Krusemeyer and Thomas [13] have criticized the artificial manner of approximating the distribution of free electrons and holes in the boundary layer shown in Fig. 5.8. They have found that the Hauffe and Engell treatment holds for depletive adsorption on n- and p-type semiconductors if the voltage drop across the space-charge layer is greater than $\approx 2kT/e$ and no degeneracy or saturation effects exist. For cumulative adsorption on either type of semiconductor, the lower limit of applicability is $\approx 4kT/e$. But the most serious fault of the theory is that it considers only those types of chemisorption which result from a transfer of charge. In conjunction with chemisorption involving charge transfer, Wolkenstein has considered simultaneous neutral adsorption.

Quantum- and Statistical-Mechanical Theories

A great deal of work has been done by scientists of the U.S.S.R. in the area of chemisorption and catalysis on semiconductors, especially by Wolkenstein [14–28]. His concepts are broader than the boundary-layer theory and include it, for he considers not only chemisorption with a charge transfer and resulting change in conductivity, but also simultaneous adsorption of neutral or polarized atoms. In fact, he postulates adsorption as C^+, C^-, and C on a single surface at the same time and calculates their equilibrium concentrations. Many examples of semiconductors that chemisorb such simple gases as hydrogen, oxygen, and carbon monoxide in different forms are known, and they support Wolkenstein's conceptions. They will be discussed in Section 5.3. Wolkenstein describes two general types of chemisorption: weak chemisorption and strong chemisorption. In weak chemisorption, there is no change in the number of electrons in the conduction band or positive holes in the valence band of the crystal. The chemisorbed particle with its adsorption center remains electrically neutral. In strong chemisorption, either donation (C^+) or capture (C^-) of an electron by the chemisorbed particle occurs, and consequently, the number of electrons in the conduction band or positive holes in the valence band of the crystal will change. The forms of strong and weak chemisorption are considered to interconvert, and Wolkenstein [28] and others [29] have developed the statistical equilibrium calculations for an intrinsic semiconductor.

Consider a total of N_s chemisorbed particles on a semiconductor surface, of which N^0 are in a state of weak or neutral bonding (C), N^- in a state of strong acceptor bonding (C^-), and N^+ in a state of strong donor bonding (C^+). The quantities η^0, η^+, and η^- are introduced, which give the probability that a chemisorbed particle will be in a given state,

$$\eta^0 = N^0/N_s, \qquad \eta^- = N^-/N_s, \qquad \eta^+ = N^+/N_s, \tag{22}$$

$$\eta^0 + \eta^- + \eta^+ = 1.$$

The probability that C^- will be formed depends on the relative positions of the Fermi level ($E_F = \mu$) of the semiconductor and the accepting level of C (electron affinity A_0 of C). We use Eq. (3) in Chapter 3,

$$\bar{n}_j = 1/\{\exp[(\epsilon_j - \mu)/kT] + 1\}, \tag{23}$$

which gives the average number of particles in a given energy state ϵ_j in Fermi–Dirac statistics. Since no more than one electron may occupy an energy state according to these statistics, $\bar{n}_j$ is numerically equal to the normalized probability. We set $\epsilon_j = A_0$ and $\mu = E_F$ to obtain

$$N^-/(N^0 + N^-) = 1/\{\exp[(A_0 - E_F/kT] + 1\} \tag{24}$$

for the probability that an accepting level of C is occupied by an electron. Note that at equilibrium the total number of electron-accepting levels of C, occupied or unoccupied, is N^0 (unoccupied) plus N^- (occupied). N^+ is not included in the total because acceptance of an electron by C^+ gives C, whose number N^0 has already been counted. Similarly, the probability that a donating level of C is unoccupied depends on the relative position of the Fermi level and the donating level (ionization potential I of C).

We obtain

$$N^+/(N^0 + N^+) = 1/\{\exp[(I - E_F)/kT] + 1\}. \tag{25}$$

With a little algebra, we find for the fraction of the atoms of C in each of the three states C, C^+, and C^-,

$$\eta^0 = 1/\{1 + \exp[(E_F - A_0)/kT] + \exp[(I - E_F)/kT]\}, \tag{26}$$

$$\eta^- = \exp[(E_F - A_0)/kT]/\{1 + \exp[(E_F - A_0)/kT] + \exp[(I - E_F)/kT]\}, \tag{27}$$

$$\eta^+ = \exp[(I - E_F)/kT]/\{1 + \exp[(E_F - A_0)/kT] + \exp[(I - E_F)kT]\}, \tag{28}$$

It is evident that the position of the Fermi level determines the ratio of the different forms of chemisorption at equilibrium. From these considerations, we see that the limitation on coverage in depletive adsorption according to the boundary-layer theory is the result of neglecting weak or neutral bonding. When weak bonding is allowed, the limitation disappears.

In another study [30], Wolkenstein discusses the dissociative chemisorption of a molecule AB on a semiconductor surface. He treats the chemisorption as a three-electron problem in quantum mechanics, postulating that as the molecule AB with a two-electron bond approaches the surface, a free electron from the conduction band is localized in the vicinity of atom B. The AB linkage is stretched, finally breaks, and a new bond between B and the lattice is formed. The conduction electron plays an essential role and it may be viewed as a free valence. These free valences are free to move in the crystal and to the surface, and their concentration is a function of temperature. The semiconductor itself is pictured as a giant "polyradical" molecule at the surface of which radical ions are formed by the process of chemisorption.

In concluding this section, we cite several other works of similar nature: the statistical calculations of Kogan and Sandomirskii for adsorption on surface defects [31] and heats of adsorption calculated from Fermi levels [32]; the statistical-mechanical calculations of Garcia–Moliner [33] for estimating the fraction of the different forms of chemisorbed particles, and the quantum-mechanical discussions of Kiselev [34].

5.3 Semiconductivity Studies and Ionic Surface Complexes

The number of papers correlating collective-electron behavior and chemisorptive phenomena is vast—over 100 papers are available on just zinc oxide and nickel oxide. The works range from empirical correlations of conductivity and chemisorption to interpretations of the types of chemisorption, to identification of surface complexes and surface states. The area cannot be treated exhaustively here. We shall endeavor to give a cross section with strong emphasis on those works of conceptual rather than empirical nature.

According to the boundary-layer theory, depletive chemisorption may occur with an electron-accepting atom on an n-type semiconductor or an electron-donating atom on a p-type semiconductor (see Table 5.1) and should be accompanied by a decrease in conductivity. Examples showing a decrease in conductivity are electron-accepting oxygen atoms on the n-type semiconductors ZnO [35], Ge [36], PbS [37], Ag_2S [38], and CdS [39]; and electron-donating carbon monoxide and hydrogen on the p-type semiconductor Cu_2O [40]. Data on coverage are scarce. At 20°C, the coverage for hydrogen on Cu_2O [41] appears to be about 5%, for CO on NiO about 1%, and CO on CoO about 3–6% [42]. But these values may not represent equilibrium coverage because the increasing potential barrier to electron transfer across the boundary layer with coverage may stop chemisorption before all electrons in the conduction band have been used. As shown in Section 5.2, the heat of adsorption should fall with increasing coverage, and this has been found true for CO on CoO [42] and CO_2 on ZnO at 450–500°C [43]. Increasing conductivity with coverage has been observed for oxygen on the p-type oxides Cu_2O, NiO, and CoO, and for hydrogen on the n-type oxide ZnO [38]. All of these systems show the high coverage characteristic of cumulative chemisorption with none of the restrictions of depletive chemisorption.

We illustrate in Fig. 5.9 the results of conductivity measurements during the chemisorption of oxygen, hydrogen, and carbon monoxide on films of p-type Cu_2O at 200°C [40]. Oxygen is chemisorbed as negative ions, indicated by the abrupt fall in resistance (rise in conductivity), which is caused by an increase in the number of current-carrying positive holes in the valence band of Cu_2O. This represents cumulative-type chemisorption, as mentioned above. On evacuation at 200°C, the initial value of the conductivity returns. On the other hand, hydrogen produces a sharp rise in resistance (fall in conductivity), which indicates the formation of positively charged ions and depletive-type chemisorption. Carbon monoxide behaves similarly, but a secondary process occurs in this case, shown by the slight reversal in conductivity change and by the fact that film resistance does not return to its initial value upon evacuation. This secondary effect is the result of carbonate ion formation, about which we shall say more later. Conductivity increases when hydrogen is chemi-

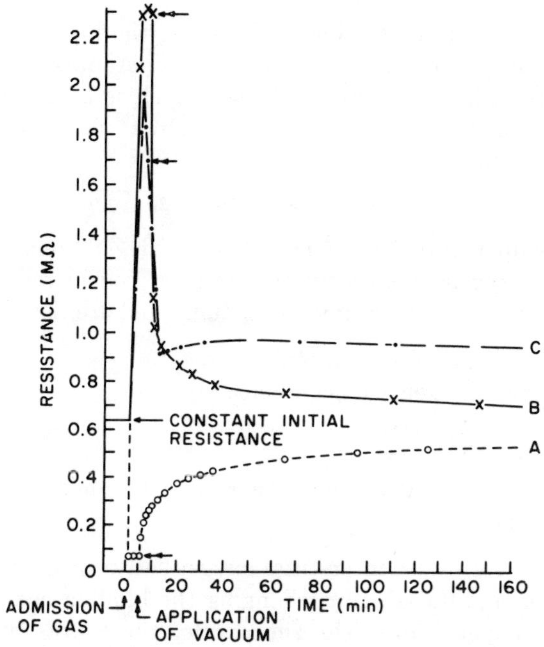

Fig. 5.9. Resistance changes during adsorption and desorption of gases on cuprous oxide at 200°C. Curve A, oxygen; curve B, hydrogen; curve C, carbon monoxide. [After Garner *et al.* [40]. By permission of The Royal Society.]

sorbed on nonstoichiometric, n-type zinc oxide (excess of Zn ions), and increases further when a higher-valent oxide, Al_2O_3, is incorporated [44].

Oxygen is always chemisorbed as a negative ion, giving a rise in conductivity for p-type oxides such as cuprous, nickel, and manganous oxides, and a drop in conductivity for n-type oxides, such as zinc oxide and titania. Conversely, hydrogen and carbon monoxide, when they affect conductivity during chemisorption, are electron donors forming positive ions.

The complexity grows when we consider that more than one form of chemisorption may take place on a particular semiconductor and that chemisorption does not always affect conductivity. The chemisorption of hydrogen on zinc oxide is an interesting case. Below 120°C, hydrogen chemisorbs readily on zinc oxide, possibly with a low heat of adsorption [38]; and when the temperature is raised above 120°C, Taylor and Liang [45] noted that hydrogen is first desorbed and then slowly readsorbed. They concluded that high- and low-temperature adsorption take place on different sites. The low-temperature adsorption occurs with no significant effect on the conductivity of zinc oxide, while the high-temperature form causes a substantial increase in conductivity. A possible interpretation [46] is that the high-temperature form, which is adsorbed irreversibly, represents a chemical combination of hydrogen atoms with oxygen ions of the zinc oxide lattice,

$$2O^{2-} + H_2 \rightarrow 2OH^- + 2e \tag{29}$$

and the low-temperature form is associated with Zn^{2+} ions.

The case of oxygen chemisorption has been considered in great detail and the various forms which have been suggested are listed as follows:

1. Physically adsorbed O_2 molecules;
2. Chemisorbed oxygen molecule ions, O_2^-;
3. Chemisorbed O^- ions;
4. Chemisorbed O^{2-} ions;
5. Oxygen chemisorbed in anion vacancies, becoming indistinguishable from lattice oxygen ions.

Other possible forms are interstitial oxygen; adsorbed oxygen into whose layer cations have diffused, thus changing the binding energy; and highly mobile lattice oxygen ions. On zinc oxide, there are two temperature maxima for the adsorption of oxygen, one at room temperature and the other just above 300°C. Adsorption is rapid and the amount is small, as would be expected in depletive chemisorption. Adsorption as O_2^- is unlikely, since the isotopic equilibrium reaction $^{18}O_2 + {}^{16}O_2 \leftrightarrows 2\,{}^{18}O^{16}O$, which requires dissociation, goes at room temperature. In both temperature regions, conductivity decreases during chemisorption. It is suggested [38] that O^- is the form predominating at room temperature, and above 300°C, O^{2-}. The high-temperature form may be stabilized by drawing interstitial zinc into the surface under the influence of adsorbed oxygen ions, as experiments on the diffusion of interstitial zinc [47] and on the sintering of zinc oxide in oxygen [48] suggest. This form is essentially incipient oxidation. A somewhat similar mechanism for adsorption of oxygen on the p-type semiconductor Cu_2O [41] has been proposed. The experimental results are explained by assuming that, in addition to forming a reactive surface layer (O^{2-}), oxygen is gradually built into the crystal by diffusion of metal ions to the surface. In this way, positive holes are formed in the interior with increase in positive hole conduction.

Carbon monoxide, like hydrogen, may adsorb both reversibly and irreversibly. At room temperature, CO chemisorption is completely reversible on ZnO [49], partly reversible and partly irreversible on Cu_2O [40] and $ZnO \cdot Cr_2O_3$ [49], and completely irreversible on Mn_2O_3 [50].

Irreversible adsorption of CO is explained by the process

$$CO + 2O^{2-} \rightarrow CO_3^{2-} + 2e, \tag{30}$$

which produces an anion vacancy containing two electrons (F' center). With this mechanism, we would anticipate subsequent oxygen adsorption to the extent of one half the CO adsorption in order to fill the anion vacancies,

$$\tfrac{1}{2}O_2 + 2e \rightarrow O^{2-} \tag{31}$$

This has been observed experimentally [40]. When CO is chemisorbed on a p-type oxide, the electrons liberated by CO will neutralize positive holes in the valence band, and on an n-type oxide will enter the conduction band. Not well understood is the change in conductivity observed during the reversible adsorption of CO on ZnO and Cu_2O.

Ingenious experiments have been devised by Morrison [51] for measuring the energy levels of the surface states of oxygen anions on a crystal of zinc oxide. He places the ZnO crystal in an oxygen atmosphere and initiates a corona discharge through the oxygen. Negative oxygen ions formed by the discharge migrate to the surface of the zinc oxide crystal. Now chemisorption of oxygen on ZnO is depletive and therefore a negative layer vastly beyond normal equilibrium for the adsorption is formed by this procedure. This depresses the Fermi level tremendously ($\sim$3eV), so that the surface states of the chemisorbed anions lie in the forbidden zone at an energy E below the bottom of the conduction band and far above the depressed Fermi level of the semiconductor. Surface charge decays by electrons moving to the interior, and this can only happen when an electron gains sufficient activation energy E to reach the conduction band levels. The quantity E measures the energy level of the surface state and is obtained by observing the rate of decay at two or more temperatures. The $(000\bar{1})$ orientation is monoenergetic at $E = 1.0$ eV, but the (0001) orientation shows a complex behavior, having three discrete levels $E_1 = 0.6$, $E_2 = 0.8$, and $E_3 = 1.0$ eV. To be more specific about surface states associated with oxygen anions, consider the singly charged ion O^-. The empty orbital in the neutral oxygen atom O, into which an electron enters to form the negative ion, is the orbital of interest. During the approach of O^- to the surface of ZnO, the energy level of the orbital is modified, becoming the surface state associated with the chemisorbed oxygen. In addition to surface states associated with chemisorbed ions, there are discrete defect states of the crystal surface itself.

Gray and Amigues [52] have employed a different technique for measuring the surface states of ZnO, which is direct and elegant. They subjected

ZnO, under well-defined conditions of pretreatment and gas atmosphere, to excitation at liquid nitrogen temperature, using a high-pressure mercury vapor discharge or a voltage-regulated strip-filament tungsten lamp and narrow band-pass interference filters. A steady electronic state is developed with a specific occupancy of all impurity levels, and this situation does not change significantly when the excitation is removed. The trapped electrons are subsequently released under slow uniform heating at the temperature corresponding to the depth of the surface energy level, and the electron current is measured by an electrometer and millivolt recorder. Measurements made both in oxygen atmosphere and *in vacuo* allow the determination of the surface states of ZnO as well as of surface states associated with oxygen. The energy levels measured were 0.12–0.15, 0.4–0.5, 0.6, 0.8, 1.0, and 1.1–1.2 eV. The first level 0.12–0.15 is identified as the defect donor level of ZnO and the level at 0.4–0.5 is believed to be associated with residual copper impurity (0.3 ppm). The level 0.8 eV, they believe, is associated with O^-, and the other levels in common with Morrison's measurements, 0.6 and 1.0 eV, are also believed to be associated with oxygen anions. The shallow levels observed by Gray and Amigues and not by Morrison are the result of starting as low as liquid nitrogen temperature, while Morrison started at ambient temperature, where the shallow levels would already have been emptied. Similar work has been reported by Gray and Cichowski [53] on titania.

The boundary-layer theory is consistent with many experimental data. In many cases, it predicts the direction of change of conductivity accompanying chemisorption, and there are examples of depletive and cumulative chemisorption, as we have seen. The effect of incorporation of foreign ions has been shown in many instances to be in accord with the theory. But there are complex situations in chemisorption on semiconductors wherein the conductivity changes are baffling or no change in conductivity occurs at all, which cannot be explained by the boundary-layer theory. We proceed to more fundamental, though less developed, concepts of chemisorption.

REFERENCES

1. N. F. Mott and R. W. Gurney, "Electronic Processes in Ionic Crystals." Oxford Univ. Press, London and New York, 1950.
2. R. A. Levy, "Principles of Solid State Physics." Academic Press, New York, 1968.
3. D. A. Wright, "Semiconductors." Methuen, London, Wiley, New York, 1950.
4. C. W. Dunlap, Jr., "An Introduction to Semiconductors." Wiley, New York, 1957.

5. D. Long, "Energy Bands in Semiconductors." Wiley (Interscience), 1968.
6. L. F. Mattheiss, *Phys. Rev.* **B5,** 290 (1972).
7. P. Aigrain and C. Dugas, *Z. Elektrochem.* **56,** 363 (1952).
8. K. Hauffe and H. J. Engell, *Z. Elektrochem.* **56,** 366 (1952).
9. H. J. Engell and K. Hauffe, *Z. Elektrochem.* **57,** 762 (1953).
10. P. B. Weisz, *J. Chem. Phys.* **20,** 1483 (1952).
11. P. B. Weisz, *J. Chem. Phys.* **21,** 1531 (1953).
12. F. S. Stone, "Chemistry of the Solid State" (W. E. Garner, ed.), p. 367. Butterworth, London and Washington D.C., 1955.
13. H. J. Krusemeyer and D. G. Thomas, *J. Phys. Chem. Solids* **4,** 78 (1958).
14. Th. Wolkenstein, *J. Phys. Chem.* (*USSR*) **22,** 311 (1948).
15. Th. Wolkenstein, *J. Phys. Chem.* (*USSR*) **26,** 1462 (1952).
16. Th. Wolkenstein, *J. Phys. Chem.* (*USSR*) **28,** 422 (1954).
17. Th. Wolkenstein and S. Z. Roginsky, *J. Phys. Chem.* (*USSR*) **29,** 485 (1955).
18. Th. Wolkenstein, *Advan. Phys. Sci.* (*USSR*) **60,** 249 (1956).
19. Th. Wolkenstein, *J. Phys. Chem.* (*USSR*) **21,** 1317 (1947).
20. Th. Wolkenstein, *Bull. Acad. Sci. USSR Div. Chem. Sci.* 916 (1957).
21. Th. Wolkenstein, *J. Chem. Phys.* **54,** 175 (1957).
22. Th. Wolkenstein, *Bull. Acad. Sci. USSR Div. Chem. Sci.* 143 (1957).
23. Th. Wolkenstein, *Bull. Acad. Sci. USSR Div. Chem. Sci.* 924 (1957).
24. Th. Wolkenstein, *J. Chem. Phys.* **54,** 181 (1957).
25. Th. Wolkenstein and V. L. Bonch-Bruevich, *J. Exp. Theoret. Phys.* (*USSR*) **20,** 624 (1950).
26. Th. Wolkenstein, *J. Exp. Theoret. Phys.* (*USSR*) **22,** 184 (1952).
27. Th. Wolkenstein, "Theorie Electronique de la Catalyse Sur les Semi-Conducteurs." Masson, Paris, 1961.
28. Th. Wolkenstein, *Advan. Catal.* **12,** 189 (1960).
29. L. I. Ahmed, *J. Phys. Chem. Solids* **29,** 1653 (1968).
30. Th. Vol'kenshtein, *Advan. Catal.* **9,** 818 (1957).
31. Sh. M. Kogan and V. B. Sandomirskii, *Izvest. Akad. Nauk, SSSR Otdel. Khim. Nauk* 1681 (1959).
32. Sh. M. Kogan and V. B. Sandomirskii, *Zh. Fiz. Khim.* **33,** 1129 (1959).
33. F. Garcia–Moliner, *Solid State Commun.* **7,** 239 (1969).
34. V. F. Kiselev, *Dokl. Akad. Nauk SSSR* **176,** 124 (1967).
35. G. Heiland, *Z. Phys.* **138,** 459 (1954).
36. E. N. Clarke, *Phys. Rev.* **95,** 284 (1954).
37. H. Hintenberger, *Z. Naturforsch.* **1,** 13 (1946).
38. F. S. Stone, "Chemisorption" (W. E. Garner, ed.), pp. 183–185. Academic Press, New York, Butterworths, London and Washington D.C., 1958.
39. P. Mark and S. Baidyaroy, *Surface Sci.* **30,** 53 (1972).
40. W. E. Garner, T. J. Gray, and F. S. Stone, *Proc. Roy. Soc.* **A197,** 294 (1949).
41. W. E. Garner, F. S. Stone, and P. F. Tiley, *Proc. Roy. Soc.* **A211,** 472 (1952).
42. R. Rudham and F. S. Stone, "Chemisorption" (W. E. Garner, ed.), p. 205. Academic Press, New York, Butterworth, London and Washington D.C., 1958.
43. T. Takaishi, *Z. Naturforsch.* **11a,** 286 (1956).
44. L. F. Heckelsberg, A. Clark, and G. C. Bailey, *J. Phys. Chem.* **60,** 559 (1956).
45. H. S. Taylor and S. C. Liang, *J. Amer. Chem. Soc.* **69,** 1306 (1947).
46. S. R. Morrison, *Advan. Catal.* **7,** 259 (1955).
47. D. G. Thomas, *Phys. Chem. Solids* **3,** 229 (1957).

48. H. J. Allsop and J. P. Roberts, *Trans. Faraday Soc.* **55,** 1386 (1959).
49. W. E. Garner and F. J. Veal, *J. Chem. Soc.* 1487 (1935).
50. T. Ward, *J. Chem. Soc.* 1244 (1947).
51. S. R. Morrison, *Surface Sci.* **13,** 85 (1969).
52. T. J. Gray and P. Amigues, *Surface Sci.* **13,** 209 (1969).
53. T. J. Gray and R. S. Cichowski, Ph.D. Thesis, N.Y. State College of Ceram. at Alfred Univ. (1968).

VI

Free-Surface States

With this chapter, we begin a more fundamental approach to the study of the chemisorptive bond. The ultimate aim is to produce a theory based on quantum-mechanical principles which describes the mutual interaction of atoms of the solid and foreign atoms on the surface. To understand in detail the methods employed requires great mathematical skill and a profound knowledge of quantum mechanics. To understand their significance requires a thorough knowledge of the experimental results of chemisorption studies. We shall treat the theoretical methods qualitatively. At this point in the development, we should not expect to find quantitative verification of chemisorption experiments. In fact, the more empirical approaches of the preceding chapters are more successful in this respect. But the depth of insight into the phenomena of chemisorption and the challenge to experiment offered by the fundamental approaches of this chapter and the following ones cannot be matched elsewhere. With danger of oversimplification, we may look upon the solid as a giant molecule with free valence at its surface that interacts with the adsorbate, forming the chemisorbed species. The term *free valence* has a nebulous quality. It might be applied reasonably to systems whose free valences are discrete levels. But in solids, particularly metals, broad bands of energy exist, and so the valence states are quasi-continuous from the ground state up. If we use the molecular orbital method, looking for one-electron wave functions of the solid plus adsorbed species as a whole, the quasi-continuous bands of energy are introduced at the outset into the system. Mathematical techniques are available for solutions of this problem based on a knowledge of the electronic structure of the adsorbed species and of the solid with a free surface, providing certain simplifying assumptions are made (see the tight binding approximation in Section 6.3). For this reason, the problem of the energy states of the free surface must be considered first.

In the first section, we review the early work of Tamm and Shockley in the 1930s. The notion of energy states, differing from those of the bulk solid and peculiar to the surface, must have been entertained long before the first paper by Tamm in 1932. The mere fact that a surface represents a discontinuity in an energy pattern is sufficient to warrant special consideration of the surface. The simple model proposed by Tamm—a one-dimensional monatomic lattice represented by a single, periodic, square-well potential—predicts a band of surface states lying outside the band of the bulk "solid." A few years later, Shockley predicted bands peculiar to the surface for a more realistic model involving more than one atomic orbital on each crystal atom. From this early period, a host of novel and refined approaches have led to a more critical defining of the conditions required for free-surface states. These are summarized in Section 6.2. The next section applies the LCAO–MO method to an idealized monatomic system, often one-dimensional, always general. Such studies represent the bulk of the work toward a fundamental approach to free-surface states and the surface states associated with chemisorption. In the last section, methods and results for heteropolar ionic systems are reviewed. The chapter ends with a brief comment on the difficulties encountered in using these fundamental methods on metallic systems.

As mentioned above, it is too early to expect exact quantitative results from these theoretical studies for real solids. But the qualitative implications of these studies are revolutionary. Furthermore, only recently have experimental methods for determining surface-state energies been developed (see Section 5.3).

6.1 Early Work

By a free-surface state, we mean the energy bands or discrete levels of the surface of a solid, with no foreign species adsorbed on it, that lie outside the region of energies of the bulk solid. Tamm [1] was the first to treat the problem of free-surface states. He considered a simple, one-dimensional, monatomic lattice and employed a modification of the

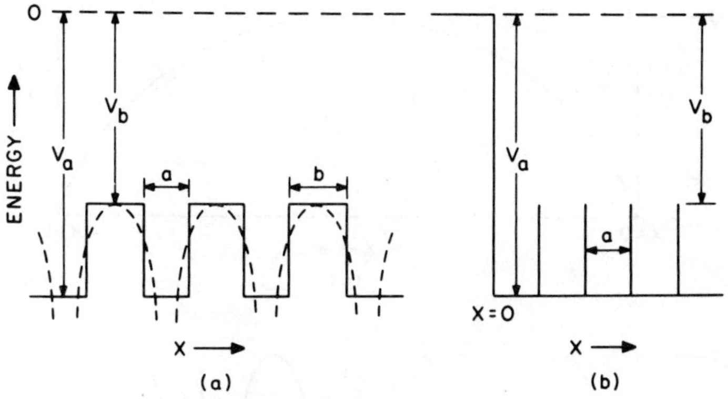

Fig. 6.1. Kronig–Penney models of the potential along a line of atoms. The dotted line shows a more realistic potential. (a) Infinite lattice ($-\infty < x < \infty$), (b) semi-infinite lattice ($0 < x < \infty$) as defined by Tamm.

quantum-mechanical model developed a year earlier by Kronig and Penney [2]. The Kronig–Penney model is shown in Fig. 6.1a, where the potential along a line of atoms is approximated by a periodic square-well potential which extends infinitely in either direction on the line x of atoms. The width a of the potential well in which an atom resides and the width b between potential wells are separated by a potential $V_b - V_a$, which is assumed to be very large, $(V_b - V_a) \to \infty$. Tamm assumed a semi-infinite lattice, that is, a line of atoms that stretches infinitely in the positive direction of x, but ends abruptly at $x = 0$, which corresponds to a surface. He further assumed infinitely narrow barriers ($b \to 0$) between the potential wells and infinitely high barriers (Fig. 6.1b). But the last well is terminated at $x = 0$ by the zero potential of the surroundings rather than by the infinite potential spike that separates adjacent wells inside the lattice. Finally, Tamm imposed a boundary condition on the wave function ψ at $x = 0$ that the propagating waves inside the lattice be smoothly joined to an exponentially decaying function outside,

$$\left(\frac{d \ln \psi}{dx}\right)_{\text{in}} = \left(\frac{d \ln \psi}{dx}\right)_{\text{out}} \tag{1}$$

where, in the three-dimensional case, x is the coordinate normal to the surface. The significance of this boundary condition may be illustrated as follows. In the free-electron model, the wave function is $\psi = \sin kx$ and represents the wave function of an electron which is restricted to one dimension and to the potential box representing the "solid," as shown schematically in Fig. 6.2a. In this case, $\psi = 0$ at $x = 0$ and $x = L$. Since the probability of an electron's being at any point is given by ψ^2, the

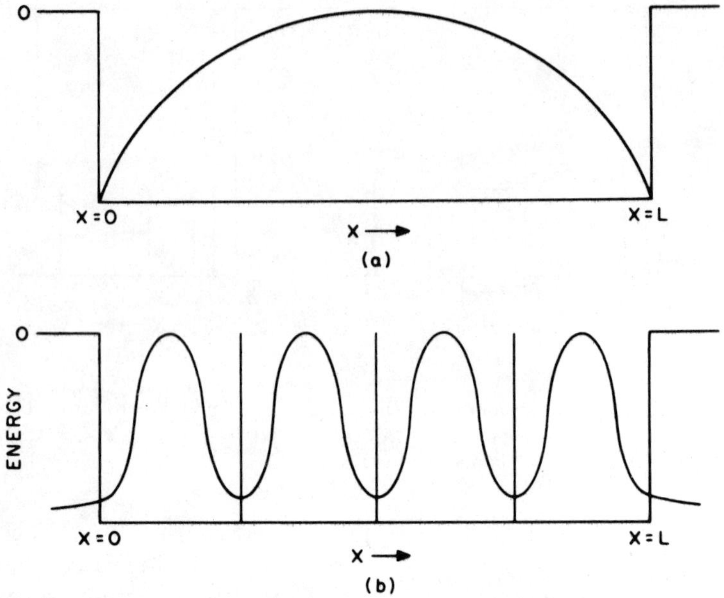

Fig. 6.2. Wave function ψ in a potential well. (a) Free-electron model with boundary conditions such that the electron is restricted to the potential box. (b) Kronig–Penney model showing the tunneling effect and finite probability of an electron be'ng found outside the potential box (lattice).

electron has zero probability of being at the boundaries of the box. In a series of such boxes, the wave functions would be identical and independent. But in the present case, Fig. 6.2a does not fit the quantum-mechanical calculations when the arbitrary restriction that an electron remain in a single potential well is lifted. These calculations indicate that there is a finite probability of the electron's being in the potential barrier. This is the well-known tunneling effect in quantum mechanics. The portions of the waves tunneling through the barriers will join together smoothly, as shown in Fig. 6.2b, and the narrower the potential barrier, the greater the probability that an electron will tunnel through it. We see that Tamm's assumption that the wave function inside the final well will be joined smoothly to the exponentially decaying wave function outside is a reasonable extension of the tunneling effect operating inside the solid.

For this model, Tamm's calculations show that it is possible to have energy levels lying outside the energy bands of the infinite line of atoms. Consequently, the wave functions are localized at the surface of the crystal. He found that one surface level was possible for each energy gap between

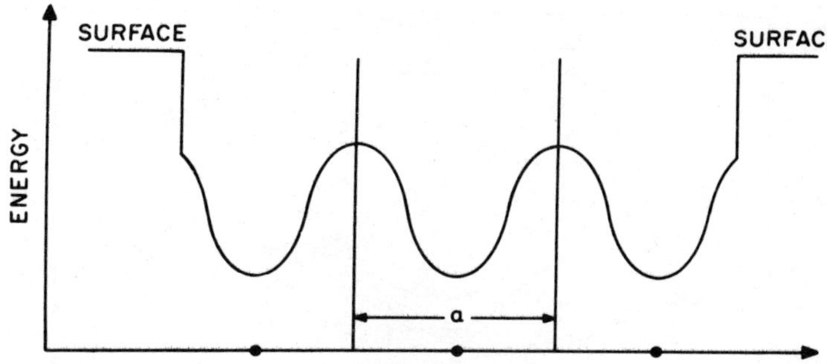

Fig. 6.3. Schematic representation of a one-dimensional monatomic lattice in the nearly-free-electron approximation, showing the lattice sites at potential minima and the lattice terminated at a potential maximum without distortion.

the ordinary allowed bands of energies. Later, Fowler [3] showed that Tamm's levels for a finite crystal should occur in pairs because of the two faces of the crystal. The spacing between the surface states becomes negligible when the opposing surfaces are more than about 10 lattice constants apart. Fowler emphasizes that these surface states are actually crystal states with wave functions extending through the lattice, though their amplitudes are negligible except near the surfaces. When the problem is extended to three-dimensional crystals, it follows that the surface levels become surface bands lying outside the crystal bands.

Goodwin [4–6] made the next major advance in 1939 when he undertook a detailed examination of the surface states of monatomic lattices by the conventional approaches of the quantum-mechanical lattice problem—the nearly-free-electron and the tight binding approximations. The tight binding approximation (see Section 2.3), a linear combination of atomic orbitals (LCAO) technique, is a perturbation technique by which the distortion of the lattice potential at the surface is included. The electrons are considered to be tightly bound to the individual potential wells located at the lattice sites. Thus, both Tamm and Goodwin assume a distortion of the potential at the surface, but Goodwin uses a more realistic periodic potential. He finds the same type of surface states as Tamm. In the nearly-free-electron approximation, the periodic lattice potential is treated as a small perturbation of the kinetic energy of the electron. For analytical reasons, Goodwin employed the nearly-free-electron approximation with negligible distortion of the potential at the surface, as illustrated in Fig. 6.3. The lattice is terminated at a potential maximum in this case instead of at a minimum as in Tamm's square-potential and Goodwin's more realistic potential in his tight binding approximation. Goodwin found that surface states were also possible from his nearly-free-electron approach

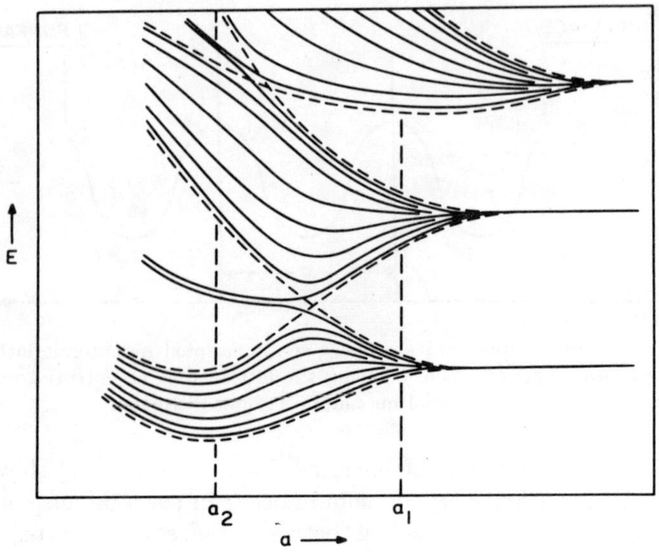

Fig. 6.4. Energy spectrum for a one-dimensional model of a lattice with eight atoms. [After Shockley [7]. By permission of the American Institute of Physics.]

with negligible distortion of the potential at the surface. But it remained for Shockley [7] to explain the physical origin of this new type of surface state.

Shockley used a model consisting of a monatomic one-dimensional crystal with a finite number of atoms, each of which is represented by a potential well whose shape is symmetrical about the center of the atom, but is otherwise arbitrary. The behavior of the potential beyond the edges of the end cells does not affect the results. The energy of the system is considered as a function of the lattice constant, as illustrated in Fig. 6.4. At large lattice constants, the energy bands, one for each atomic orbital per crystal atom, are narrow, since there is small interatomic overlapping of the orbitals. Surface states first appear after the boundary curves of the energy bands have crossed. After they have crossed, two energy levels have separated, one from each of the bands. We note that such surface energy levels result not only from the assumption of negligible distortion of potential at the surface, but also from the use of more than one atomic orbital on each crystal atom. In the three-dimensional case, the surface levels are broadened into surface bands.

Shockley states may be considered to be the free valencies at the surface in the usual sense. For example, the four valence electrons of carbon occupy four atomic orbitals in the isolated state of the atoms, one s orbital and three p orbitals. In bonding with other atoms, these valence electrons are hybridized into sp^3 wave functions giving tetrahedral coordination. Including spin, there are eight states, four occupied in bonding and four in antibonding orbitals of higher energy. In the diamond crystal structure, the discrete levels are broadened into bonding and antibonding bands. On the (111) surface, three orbitals are used in bonding the surface atom to the crystal; the fourth is free. This "dangling" orbital localized on the surface is the Shockley state and may form surface bands.

In summary, the early work showed that surface states are possible under the following conditions:

1. the use of one or more atomic orbitals per crystal atom; distortion of the potential at the surface (corroborated by the calculations of Tamm for a square potential and by the tight binding, quantum-mechanical calculations of Goodwin for one atomic orbital per crystal atom);
2. the use of more than one atomic orbital per crystal atom; negligible distortion of the potential at the surface (corroborated by the work of Shockley and the nearly-free-electron calculations of Goodwin).

It has become customary to refer to these two types of surface states as Tamm states and Shockley states, respectively. Calculations for Tamm states using the LCAO–MO procedure will be illustrated in Section 6.3. The corresponding problem for Shockley states is difficult to treat even by the simplest LCAO approximation.

6.2 Summary of Theoretical Techniques for the Study of Surface States

Although a great number of techniques have been developed for calculating surface states, they may be grouped into a small number of categories. Again, within each category, the number of major approaches can be reduced to a few, but below these the number of minor variations is legion.

Most of the theoretical work on determining the energy states of a finite or semi-infinite lattice is aimed at solving the Schrödinger wave equation. Two basic approaches have been widely used: (1) the potential method and (2) the LCAO method. We shall consider the potential method first.

In the potential method, an electron potential curve is assumed, including its shape at the surface. Using these potentials in the wave equation, solutions may be obtained which lead to characterization of the energy states. The calculations have been carried out in detail for one-dimensional crystals for a variety of assumed potentials. We have mentioned two types of assumed potentials in Section 6.1—the modified Kronig–Penney model employed by Tamm [1] and the nearly-free-electron model of Goodwin [4, 5, 6]. We described the modified Kronig–Penney model (Fig. 6.1) as a square potential with infinitely narrow barriers between potential wells and with the barrier at the surface rising from the bottom of the well to zero potential of the surroundings. This model is sometimes called a δ-function repulsive potential, since a δ function is by definition represented by a vertical line. Another type of assumed potential uses a δ-function attractive potential [8], in which the width of the potential well is narrowed to a vertical spike, while the distance between wells remains finite. In the nearly-free-electron approach of Goodwin, as we have mentioned, a periodic potential with no edge effects is used (Fig. 6.3). A fourth type of potential is the sinusoidal potential with variable amplitude used by Levine [9], and it is especially important for dealing with heteropolar ionic crystals. The potential is selected to be one-dimensional and sinusoidal with a variable amplitude which measures the "ionicity." The crystal potential is terminated at an arbitrary position in the last cell and with arbitrary step height. Inside the crystal, the Schrödinger equation reduces in this case to the Mathieu differential equation, whose properties are well known. The calculations show that ionic states appear in the appropriate band gap. For example, when a one-dimensional lattice is terminated at an M^+ ion, it has an M-like surface state that moves closer to the conduction band the greater the ionicity. Correspondingly, when the lattice is terminated at an X^- ion, it has an X-like surface state that moves closer to the valence band the greater the ionicity.

The LCAO method brings up fewer perplexing problems on the form of the potential at the surface; computational problems are simpler, and the method is more generally useful, especially in problems of chemisorption and catalysis. As described in Section 2.3, the wave function ψ of the finite or semi-infinite crystal is represented by a linear combination of atomic orbitals ϕ_r,

$$\psi = \sum_r C_r \phi_r. \tag{2}$$

There is wide latitude in the choice of atomic orbitals, as indicated in standard texts on quantum theory. In many cases [10], the hydrogen s orbital is used for successive atoms on a line, so that the system is really a one-dimensional hydrogen crystal. Wannier functions [11], Green's function [12], and Bloch orbitals [13] have also been used in conjunction with the Schrödinger equation. After selecting the atomic orbitals, the wave function is incorporated into the wave equation,

$$H\psi = E\psi, \tag{3}$$

where H is the one-electron Hamiltonian operator, usually without any electron–electron repulsive potential even in the sense of a self-consistent field, and E is the orbital energy. Multiplying by ψ and integrating, we obtain

$$E = \int \psi H \psi \, d\tau \Big/ \int \psi^2 \, d\tau. \tag{4}$$

The integrals in Eq. (4) contain the following specific integrals:

$$\int \phi_r \phi_s \, d\tau, \qquad r, s = 0, 1, 2, \ldots, \text{ overlap integral;} \tag{5}$$

$$\int \phi_r H \phi_s \, d\tau = H_{rs}, \qquad r \neq s, \text{ resonance integral;} \tag{6}$$

$$\int \phi_r H \phi_r \, d\tau = H_{rr}, \qquad \text{Coulomb integral.} \tag{7}$$

The integral $\int \phi_r^2 \, d\tau$ is usually normalized for each r. The energy is obtained by the variation method, in which the energy is minimized with respect to each constant coefficient C_r, that is, $(\partial E/\partial C_r) = 0$. This gives a set of secular equations which may be put into determinantal form,

$$\mathbf{QC} = 0, \tag{8}$$

where $\mathbf{Q}$ is an $N \times N$ matrix (for N atoms) whose terms are the integrals of Eqs. (5), (6), and (7), and $\mathbf{C}$ is a column matrix of the constant coefficients C_r. Various simplifications in $\mathbf{Q}$ are now usually made. For example, in the tight binding approximation, which is the most common one, the overlap integral, Eq. (5), is set equal to zero; the resonance integral, Eq. (6), is restricted to nearest neighbors, $H_{r,r\pm1}$. Selection of the character of the potential at the surface is replaced in the LCAO method by assigning values to the Coulomb integral (H_{rr}) and resonance integral (H_{rs}). For example, the surface Coulomb integral (H_{00}) is usually different from the interior Coulomb integrals and is considered a "surface perturbation." The surface resonance integral H_{01} may also be different from the interior resonance integrals and is considered a "deformation." Replacement of a potential discontinuity at the surface by unique Coulomb or resonance integrals for the surface atom is conceptually clearer than characterizing surface potentials. Solution of Eq. (8) gives the energy states, both surface and bulk, when all appropriate surface and bulk Coulomb and resonance integrals are included in $\mathbf{Q}$. For the simplest problem, which is the linear one, N is the number of atoms in the crystal.

When there is more than one dimension, orbital, band, or atom per unit cell, Eq. (8) still applies, though N is not the number of atoms in the crystal any more. In the final step, the solution of Eq. (8), four methods have been used: (1) the coefficient method; (2) the scattering amplitude method; (3) the determinant method; and (4) the integration method. The coefficient and the scattering amplitude methods involve making correct guesses on the nature of $\mathbf{C}$; the determinant method endeavors to solve the determinant of $\mathbf{Q}$ ($|\mathbf{Q}| = 0$) directly, ignoring $\mathbf{C}$; and the integration method considers the unperturbed problem $\mathbf{Q}^0\mathbf{C}^0 = 0$ and then introduces the surface perturbation by an integral method. For our purposes, we need not go into any more detail.

We conclude this section with a discussion of two methods of calculating surface states that do not fit easily into the two broad categories discussed above. The free-electron network (FEN) theory has been applied recently to the calculation of surface states [14]. Originally, the theory was developed for molecules and was used especially in the study of conjugated systems of π double bonds. In the theory, the motion of π electrons is restricted to narrow channels (the branches of the network) joining centers of bonded atoms. Within the channels, the effective potential is taken as zero, and elsewhere as infinite. The Schrödinger equation is solved for one-dimensional waves propagated along the branches and eigenvalues determined for boundary conditions. Application of the theory to periodic structures was suggested by Coulson [15]. He showed that for electrons in graphite the FEN and LCAO methods are in agreement. In fact, he concludes that many features of energy bands in crystals are to a large degree independent of the theoretical model chosen. Hoerni [16] worked out the general FEN theory for an arbitrary network. A subsequent paper by Scherer and Phariseau [14] applied FEN theory to surface problems. They pictured the surface as a regular (111) crystal plane, each atom bound to three neighbors in the second atomic layer by sp^3 hybrid bonds. The fourth bond dangles perpendicular to the surface. A similarity between the Kronig–Penney potential and the potential along a branch in the FEN theory can be seen. The potentials V_a and V_b in the Kronig–Penney model (Fig. 6.1) correspond to the branch potentials 0 and ∞ in the FEN model.

Another method, more classical in nature, is based on Madelung sums and classical electrostatics, and is applicable to heteropolar ionic lattices [17]. The bulk ions are assumed to be classical point ions with fractional charge. The surface ions are assumed to be equivalent to bulk ions, except that their Madelung energies (sum of the electrostatic energies acting on an ion by the other ions of the lattice) are less because the Coulomb three-dimensional sums are only over half-space. Thus M and X pairs of surface states should appear. The M (or X) states arise when there is an M (or X) ion on the surface. Surface-state energies in the range of a few tenths of an eV have been calculated from data on fractional ionic charge, first and second ionization potentials and electron affinities, internuclear spacing, crystal bulk type, and crystallographic direction. The relationship of the model to the LCAO solution of the Schrödinger equation is not clear. (An LCAO approach to ionic crystals will be discussed in Section 6.4.)

6.3 The General Monatomic Crystal

We shall describe the calculation of surface and bulk energy states by the LCAO techniquc, tight binding approximation, for the monatomic one-dimensional system shown in Fig. 6.5 [11, 18, 19]. The two end atoms, 0 and N, represent the free "surfaces." This model possesses all the essential features of the problem. Following Grimley [10], we illustrate the principles involved. Each atom r is associated with one atomic orbital ϕ_r which need not be specified in this general case. As in Section 6.2, we form the wave function of the system ψ by a linear combination of these functions with the constants C_r,

$$\psi = \sum_r C_r \phi_r. \tag{9}$$

Using the one-electron Hamiltonian operator H for the chain satisfies the wave equation

$$H\psi = E\psi, \tag{10}$$

which may be put into the variational form

$$E = \int \psi H \psi \, d\tau \bigg/ \int \psi^2 \, d\tau. \tag{11}$$

Substituting Eq. (9) into Eq. (11) and minimizing the energy according to the variational procedure, $[(\partial E/\partial C_r) = 0;\ r = 0, 1, \ldots, N]$, gives the usual set of N linear equations for the unknown wave function co-

Fig. 6.5. One-dimensional model of a monatomic crystal.

efficients C_r and energies E,

$$(E - H_{rr})C_r = \sum_{s(\neq r)} H_{rs}C_s, \qquad r = 0, 1 \ldots, N, \tag{12}$$

where

$$H_{rs} = \int \phi_r H \phi_s \, d\tau \tag{13}$$

is the resonance integral between atoms r and s and H_{rr}, the Coulomb integral. Each of these integrals is integrated over all space and has a constant value depending on the values of ϕ. In Eq. (12), the overlap integrals of the form $\int \phi_r \phi_s \, d\tau$, $r \neq s$, which are obtained on substituting Eq. (9) into Eq. (11), have been set equal to zero, as stipulated in the tight binding approximation. Also, integrals of the form $\int \phi_r^2 \, d\tau$ are normalized. Further, in accordance with the tight binding approximation, we include the resonance integral only between nearest neighbors, $H_{r,r\pm1}$. We set

$$H_{r,r\pm1} = \beta; \qquad H_{rr} = \alpha, \qquad r \neq 0; \qquad \text{and} \qquad H_{00} = \alpha'; \tag{14}$$

and so recognize the existence of a free surface by making the Coulomb integral α' on the end atom, $r = 0$, and α otherwise. This corresponds to a distortion of the potential at the surface in the potential method (Section 6.2). Equation (12) now becomes

$$(E - \alpha)C_r = \beta(C_{r+1} + C_{r-1}), \qquad r \neq 0, \tag{15}$$

with the boundary condition

$$(E - \alpha')C_0 = \beta C_1, \qquad r = 0. \tag{16}$$

A boundary condition is needed for the other end of the chain, $r = N$. But if N is large, this boundary condition cannot affect what happens at the other end, $r = 0$, so we set $C_N = 0$. One of the methods mentioned in Section 6.2 for solving the wave equation was the coefficient method, in which correct guesses on the nature of the C_r are made. We use this method here, letting

$$C_r = \sin(N - r)\theta, \qquad r = 0, 1, \ldots, N. \tag{17}$$

The boundary condition at $r = N$ is satisfied by Eq. (17), and Eq. (17) satisfies Eq. (15) if

$$E = \alpha + 2\beta \cos \theta, \tag{18}$$

as may be shown by substitution and a little trigonometry. Finally, Eq.

(17) satisfies Eq. (16), the boundary condition at $r = 0$, if θ is one of the N roots of

$$z + \cos\theta + \sin\theta \cot N\theta = 0 \tag{19}$$

where

$$z = (\alpha - \alpha')/\beta. \tag{20}$$

Therefore, Eq. (17) is a solution to Eq. (15) if E and θ are expressed by Eqs. (18) and (19), respectively. We need to analyze the roots of Eq. (19) in order to determine the energy states. An analysis shows that the equation has $(N - 1)$ real roots. The wave functions that correspond to Eq. (17) are periodic, and if we write $E' = (E - \alpha)/2\beta$, we find from Eq. (18) that the energy levels lie in the range

$$-1 < E' < 1. \tag{21}$$

Note that E and E' are linearly related, since α and β are constants. This expression defines a band of nonlocalized states of width 4β, and all of the levels in this band arise from a single state ϕ of the isolated atoms. If z lies between -1 and $+1$, then the remaining root will also be real and represent energies lying within the band. The one-dimensional monatomic crystal has only nonlocalized states in this case, which means that all energy states, surface and bulk, lie within one quasi-continuous band. But if z is greater than $+1$ or less than -1, then the remaining root is complex and of the form $\theta = i\xi$ or $\theta = \pi + i\xi$, where ξ is real and positive. These states, associated with the free end $r = 0$, are localized, and their energies lie outside the band of levels given by Eq. (21). When $\theta = i\xi$, E' is positive and is called a $\mathcal{P}$ state; when $\theta = \pi + i\xi$, E' is negative and is called an $\mathfrak{N}$ state. Thus, from Eq. (19), a $\mathcal{P}$ state exists if

$$-z = \cosh\xi + \sinh\xi \coth N\xi, \tag{22}$$

where relations such as

$$(e^{i\theta} + e^{-i\theta})/2 = \cos\theta = (e^{\xi} + e^{-\xi})/2 = \cosh\xi \tag{23}$$

have been used. When N is very large,

$$-z = e^{\xi} \qquad (\xi > 0), \tag{24}$$

and a $\mathcal{P}$ state exists if $z < -1$ with wave function and energy given by

$$C_r = C_0 e^{-r\xi}, \tag{25}$$

$$E = \alpha + 2\beta \cosh\xi. \tag{26}$$

In a similar manner, $z = e^{\xi}$ for an $\mathfrak{N}$ state when N is very large, and such

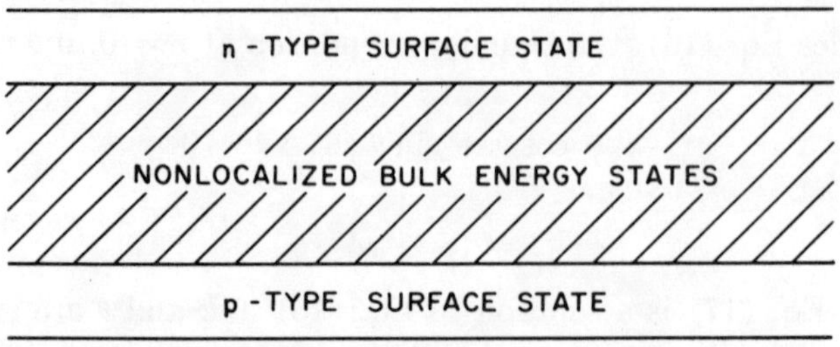

Fig. 6.6. Energy states, surface and bulk, for a one-dimensional monatomic crystal.

a state exists if $z > 1$, with wave function and energy given by

$$C_r = C_0(-1)e^{-r\xi}, \tag{27}$$

$$E = \alpha - 2\beta \cosh \xi. \tag{28}$$

We summarize the situation for a particular case. Assume $\beta < 0$, which is true when the atomic orbitals ϕ have the symmetry of atomic s states. When $z = 0$, the chain of N atoms has N nonlocalized states with a band width of 4β. As z decreases (end atom electron attracting), all of these states decrease in energy. At $z = -1$, a $\mathcal{P}$-type end state just separates below the band, and continues to drop with further decrease in z. Its wave function concentrates more and more on the end atom. In other words, the further the surface energy state drops below the bulk energy band, the more time the electron will spend on the end atom and the less time on the other atoms. As z approaches $-\infty$, the $\mathcal{P}$ state becomes the completely discrete atomic orbital ϕ_0, and the wave functions for the nonlocalized states are zero on the end atom. The situation is analogous for $\mathfrak{N}$ states as z increases above zero (end atom electron repelling), but the level separates above the band instead of below. The energy states for surface and bulk are illustrated in Fig. 6.6.

The three-dimensional case is in principle a straightforward extension with one difference: the discrete $\mathfrak{N}$ and $\mathcal{P}$ end states of the one-dimensional case for $z > 1$ and $z < -1$ now broaden into bands of surface states of width 8β containing N^2 levels. As z approaches ∞ or $-\infty$, the surface bands separate completely from the normal crystal bands and the wave functions of the normal crystal bands are vanishingly small on the surface atoms. Since these states were derived from one atomic orbital per crystal atom, and with a different energy parameter (Coulomb integral) for the surface atoms, they are Tamm states. We note that Tamm states may also be obtained by using a different resonance integral, β', for the surface atoms. The atoms of real crystals have more than one band. Metals, as we have seen, have a complex structure of overlapping bands. If more than one atomic orbital per crystal atom is used in energy-state calculations, surface states may be obtained even though $\alpha = \alpha'$ and $\beta = \beta'$. These are the Shockley states which were discussed in Section 6.1, and they are difficult to treat even in the simple LCAO tight binding approximation. For critical reviews of LCAO techniques applied to surface problems, the reader is referred to the publications of Levine and Mark [20], Koutecky [21], and Davison and Levine [22].

Extensive studies continue in an effort to bring more realistic features into the general case. Kalkenstein and Soven [12] have carried out extensive LCAO-type calculations for the general cubic lattice, (100) and (111) faces. Their method differs from other LCAO tight binding techniques in that they employ a Green's function formalism to investigate the density of energy states in semi-infinite crystals, thus giving a more complete picture of surface and bulk conditions. Their study includes the effects of altering the surface potential. It is limited to the use of one atomic orbital per crystal atom and thus to a single bulk energy band, but should not be difficult to extend to the more realistic multiband situation. A result of great interest, which they discuss in detail, is that the number of surface states in a surface band is not necessarily governed by the number of surface atoms. Neither are these surface states always severely localized, but on the contrary may extend significantly into the crystal, corresponding to the foregoing linear case with z not too far above $+1$ or below -1. The authors believe these effects are closer to real systems.

Simultaneously, studies directed toward specific types of solids, such as ionic and metallic crystals, are going on, and these will be discussed in the next section.

6.4 The Heteropolar Ionic Crystal

In the last section, the loci of atoms in the surface network were considered to be an exact replica of the bulk arrangement. Energy perturbations on the surface were assumed to be wholly the result of the lattice discontinuity. As we come closer to real systems, the possibilities of distorted surfaces, where the surface atoms have moved short distances without breaking bonds, and reconstructed surfaces, where surface bonds are broken and surface atoms have migrated to new sites, must be admitted. No examples of the reconstruction of the surfaces of heteropolar ionic crystals are known [23]. Distortions *normal* to the surface of many ionic crystals have been demonstrated, and data for alkali halides are given in Table 6.1. But *lateral* distortions are far less common. In fact,

TABLE 6.1

Equilibrium Configuration of Outermost Atoms of the (100) Face of Alkali Halide Crystals[a]

	Cation	Anion		Cation	Anion
NaF	−0.0393	−0.0217	KI	−0.0647	0.0229
NaCl	−0.0871	0.0356	RbF	−0.0054	−0.0390
NaBr	−0.1147	0.0727	RbCl	−0.0243	−0.0102
NaI	−0.1395	0.0947	RbBr	−0.0318	−0.0024
KF	−0.0133	−0.0335	RbI	−0.0455	−0.0098
KCl	−0.0335	−0.0047	CsF	−0.0091	−0.0656
KBr	0.0465	0.0021	LiF	−0.0697	0.0014

[a] From Benson *et al.* [24]. Displacements normal to the ideal (100) plane are in units of nearest-neighbor distance; negative values are displacements into the solid.

the ionic surfaces of MX crystals such as (110) zinc blende, $(11\bar{2}0)$ wurtzite, and (100) NaCl have a lateral periodicity identical to the bulk. Apparently, the lateral electrostatic forces stabilize these surfaces against distortion or reconstruction, even though the M and X ions usually have only a fractional charge. Comparing surface-state theory and experiment for these ionic surfaces appears more hopeful than for reconstructed surfaces.

A simple argument reveals interesting possibilities for surface states of ionic crystals [10, 25, 26]. We assume that all of the perturbations introduced by a free surface are associated with the change in electrostatic environment of an ion in going from the interior to the surface. Further, we assume that the normally filled valence band is associated with the anions, which is true for alkali halides and certain n-type semiconducting oxides. For this situation, the surface perturbation acts in the direction of creating a band of surface states with its center lying *above* the center of the normal anion band (valence band). This anion surface band will normally be completely filled. Conversely, the normally empty cation band (conduction band) is perturbed at the surface in the direction of creating a surface cation band with its center lying *below* the normal cation band. Thus there is a narrower band gap between filled and vacant bands at the surface than in the bulk. If the surface anion and cation bands move sufficiently to overlap, intrinsic surface conductivity or even metallic-like surface conductivity could occur. Such effects are important in the theory of chemisorption. But this simple picture must not be pushed too far; for example, it is not an adequate picture for transition metal oxides. We turn to the detailed approaches to ionic surface states.

Three theoretical methods find application in calculating surface states of ionic crystals: (1) the Madelung potential method and (2) the sinusoidal potential method, both of which are mentioned specifically in Section 6.2; and (3) the band-edge method, which involves a modified LCAO technique.

The Madelung Potential Method†

The Madelung potential method is a direct extension of the classical theory of bulk states of ionic lattices. We may speak of bulk conduction bands and valence bands of ionic crystals associated with the M^+ and X^- species in connection with this method. But it must be remembered that classical methods do not give band broadening, only discrete levels. We may think of the discrete upper and lower levels associated with M^+ and X^- as defining a band gap in which may be found the corresponding levels (or bands) of the surface states. What we seek are the changes in potential on introducing an electron into the bulk crystal and into the surface layer, and the problem can be solved by going through the following cycle:

1. Remove an M^+ ion from the crystal; the change in potential is $C_M Ze/r_0$.
2. Bring an electron from infinity to the normal state in this ion to form the atom M; the potential change is $-I_M$, where I_M is the ionization potential of the atom M.
3. Bring atom M back to the place vacated by the ion M^+. No potential change.

In the first operation, Z is the valence of the ion and may be fractional to account for covalent contributions to binding, e is the electronic charge, and r_0 is the lattice constant. The Madelung constant C_M measures the summation of the interaction of an ion with all the other ions of the crystal. Its value depends on the lattice geometry and it is essentially a correction factor that tells how much of an error will be made if only nearest neighbors are considered. It is computed for M^+ from the lattice sum

$$C_M = \sum_{i,j,k} q_{i,j,k}/R_{i,j,k} = \sum_{i,j,k} Q \tag{29}$$

where $q_{i,j,k} = \pm 1$, depending on the sign of the ion located at the index position (i, j, k), and $R_{i,j,k}$ is the distance measured from $(0, 0, 0)$ to

† Ref. [17].

(i, j, k) expressed in units of the nearest cation–anion distance. Each index runs from $-\infty$ to $+\infty$. In the classical limit, then, we see that the potential change on introducing an electron into the cationic level (corresponding to the bulk conduction band in quantum theory) is the sum of the three steps in the cycle just described,

$$E_{\mathrm{M}} = (C_{\mathrm{M}}Ze/r_0) - I_{\mathrm{M}}. \tag{30}$$

Similarly, for the negative ion X^-, we get the potential change on introducing an electron into the anionic level (corresponding to the bulk valence band in quantum theory),

$$E_{\mathrm{X}} = -(C_{\mathrm{X}}Ze/r_0) - A_{\mathrm{X}}, \tag{31}$$

where A_{X} is the electron affinity of X. Equations (30) and (31) represent the discrete bulk electronic states of the crystal, and relate to the energy levels of the bottom of the conduction band and the top of the valence band, respectively, in quantum-mechanical terms. If the lattice contains an equal number of positive and negative ions, as in M^+X^-, the Madelung constants are equal, $C_{\mathrm{X}} = C_{\mathrm{M}} = C$.

Corresponding energy states may be calculated for the surface, where the Madelung constant becomes

$$C_{\mathrm{s}} = \sum_{i\geq 0, j, k} Q, \tag{32}$$

which is the sum over half-space, the index i being normal to the surface plane $i = 0$, and j and k projecting parallel to the surface plane. Equation (32) is correct only when it is assumed that there is no rearrangement of the surface lattice and no significant change in the lattice constant r_0 at the surface. The computation of C_{s} is facilitated by rewriting Eq. (29),

$$C = \sum_{i<0} Q + \sum_{i=0} Q + \sum_{i>0} Q = \sum_{i=0} Q + 2\sum_{i>0} Q \tag{33}$$

where the second equality is the result of symmetry above and below the surface plane $i = 0$. In all summations, it is understood that the indices j and k extend over all space ($+\infty$ to $-\infty$). We may also rewrite Eq. (32) as

$$C_{\mathrm{s}} = \sum_{i=0} Q + \sum_{i>0} Q, \tag{34}$$

and combining Eqs. (33) and (34), we obtain

$$C_{\mathrm{s}} = \tfrac{1}{2}C + \tfrac{1}{2}\sum_{i=0} Q. \tag{35}$$

Since C is known, C_{s} may be computed by evaluating $\frac{1}{2}\sum_{i=0} Q$, which is the Madelung constant for the surface plane (sum over 0, j, k). Surface energy states, cationic and anionic, may now be obtained by using Eqs. (30) and (31) with C_{M} and C_{X} replaced by C_{Ms} and C_{Xs}, remembering that $C_{\mathrm{Ms}} = C_{\mathrm{Xs}} = C_{\mathrm{s}}$ for a surface containing an equal number of positive and negative ions.

$$E_{\mathrm{Ms}} = (C_{\mathrm{Ms}}Ze/r_0) - I_{\mathrm{M}}, \tag{36}$$

$$E_{\mathrm{Xs}} = -(C_{\mathrm{Xs}}Ze/r_0) - A_{\mathrm{X}}. \tag{37}$$

The difference in the bulk energy levels is

$$E_{\mathrm{M}} - E_{\mathrm{X}} = (C_{\mathrm{M}}Ze/r_0) + (C_{\mathrm{X}}Ze/r_0) - I_{\mathrm{M}} + A_{\mathrm{X}}, \tag{38}$$

with a corresponding equation for surface levels.

To get a clearer physical picture of the formation of surface states, let us plot E_{M}, E_{Ms}, E_{X}, E_{Xs} as functions of r, the lattice constant, for Eqs. (30), (31), (36), and (37), starting with the infinitely dilated lattice $r = \infty$. Referring to Fig. 6.7, we see that at $r = \infty$,

$$E_{\mathrm{Ms}}(\infty) - E_{\mathrm{Xs}}(\infty) = E_{\mathrm{M}}(\infty) - E_{\mathrm{X}}(\infty) = E(\infty) = -I_{\mathrm{M}} + A_{\mathrm{X}}. \tag{39}$$

As r decreases, the cation and anion energy curves approach each other until finally they cross, $\Delta E = 0$. The surface states have already separated from the corresponding bulk states and $\Delta E_{\mathrm{s}} < \Delta E$ to the left of the cross point where the ionic configuration is stable. In a quantum-mechanical calculation, these states would broaden into bands to the left of the cross point.

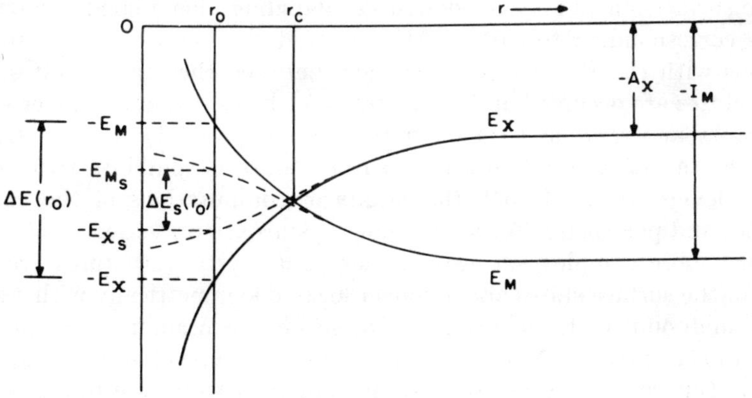

Fig. 6.7. Change in electronic energy states of an ionic lattice M^+X^- with contraction of the lattice from infinite dilation. To the left of the cross point, the ionic configuration is stable. Surface states shown by dashed lines.

TABLE 6.2

Values of Energies and Parameters Associated with Ionic Lattices[a]

Substance	Lattice type	C_s/C [b]	$\Delta E(r_0)$ [c] (eV)	Surface index	Z [d]	$\Delta E_s(r_0)/\Delta E(r_0)$ [e]	ΔE_t [f] (eV)
CdO	NaCl	0.96	2.5	(100)	2	0.9	0.13
					1	0.94	0.075
					0.5	0.95	0.063
ZnO	Wurtzite	0.88	3.2	$(11\bar{2}0)$	2	0.73	0.44
					1	0.83	0.28
					0.5	0.86	0.24
CdS	Wurtzite	0.88	2.5	$(11\bar{2}0)$	2	0.71	0.35
					1	0.80	0.25
					0.5	0.82	0.22
ZnS	Zinc blende	0.85	3.7	(110)	2	0.65	0.63
					1	0.75	0.46
					0.5	0.77	0.40

[a] Mark [17].
[b] Ratio of Madelung constants, surface/bulk.
[c] Bulk energy band gap width.
[d] Valence.
[e] Ratio of surface energy band gap to bulk energy band gap.
[f] Energy gap between upper intrinsic surface state and bottom of the conduction band (upper intrinsic bulk state). The lower intrinsic surface state is located symmetrically with respect to the midpoint of the energy gap $\Delta E(r_0)$ for this simple model.

The surface states may be viewed as "dangling" ionic bonds, the upper states corresponding to surface M^+ ions and the lower states to surface X^- ions with the Fermi level in the gap between the surface states. The lower states are occupied and act as traps for holes, while the upper states are empty and serve as electron traps (see Chapter 5.1). If the lattice is finite instead of semi-infinite, each surface state is split into two closely spaced levels. In Table 6.2, the results of computations of the various energies and parameters for a few ionic crystals are presented.

When more complex surfaces, planes, and crystal structures are considered, the surface states are no longer located symmetrically with respect to the midpoint of the energy gap. For surfaces containing only one type of ion [(111) plane of NaCl structure], only one type of surface state will appear. But such surfaces are unstable and tend to buckle to relieve the strong electric field. If instead of the point-ion approximation, the real dimensions of ions are used, C_M and C_X as well as C_{Ms} and C_{Xs} are no longer equal and the surface levels are no longer symmetrically located relative to the middle of the gap. Vacancies, edges, corners, and other geometrical irregularities which are present in real crystals may be treated in principle by the present theory, but will introduce additional surface states.

The Sinusoidal Potential Method with Variable Amplitude†

In this method of studying surface and bulk states, a one-dimensional model is used with a sinusoidal potential solved for arbitrary values of the sinusoidal amplitude q, lattice termination at z_0, surface step height ξ, and band gap index m. Inside the crystal, the Schrödinger equation turns out to be equivalent to the Mathieu equation, whose wave function and eigenstates are known. The surface states are constructed as hybrids of the bulk states, and they move toward the centers of the band gaps as in the Madelung potential method. Since the sinusoidal method is a quantum-mechanical approach, the bulk states broaden into bands and in the three-dimensional case the surface states would also broaden.

If, in the Schrödinger equation,

$$\frac{\hbar^2}{2m_e}\frac{d^2\psi(x)}{dx^2} + \left(E - V_0 - V_s \cos\frac{2\pi x}{r_0}\right)\psi(x) = 0, \tag{40}$$

where r_0 is the lattice constant, V_s is the sinusoidal potential amplitude, and V_0 is the reference potential, we make the following substitutions,

$$z = \pi x/r_0, \qquad T = \hbar^2/8m_e r_0^2, \qquad \alpha = (E - V_0)/T, \qquad 2q = V_s/T, \tag{41}$$

we get the Mathieu equation

$$\frac{d^2\psi}{dz^2} + (\alpha - 2q\cos 2z)\psi = 0, \tag{42}$$

whose wave functions ψ and energy states $\alpha = (E - V_0)/T$ are known. The surface is located at an arbitrary point z_0 which may or may not be at a wave maximum ($z_0 = 0$) or wave minimum ($z_0 = \frac{1}{2}\pi$) of the unit cell. The crystal is assumed infinite for $z \geq z_0$. Inside the crystal, the potential is sinusoidal; outside the crystal, $z \leq z_0$, the potential has a step height V_0/T. Finite crystals are also considered. Since Eq. (42) is solved for all appropriate wave functions ϕ and corresponding energies α, there will be a series of energy bands and band gaps, and the band gaps are enumerated $m = 1, 2, \ldots$. The surface states—all Tamm states—can be made to correspond to metallic, covalent, and ionic states on real crystals, in accord with LEED. Figure 6.8 shows the bulk energy bands and gaps

† Ref. [9].

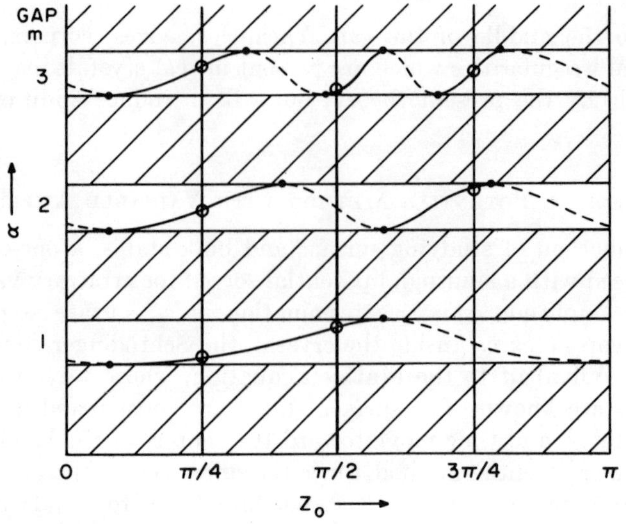

Fig. 6.8. Surface-state energies versus z_0 for a strongly periodic lattice and finite surface potential. Covalent surface states are in gaps m = odd; ionic surface states in gap $m = 2$. [After Levine [9]. By perm'ssion of the American Institute of Physics.]

($m = 1, 2, 3$) of energies $\alpha = (E - V_0)/T$ plotted as a function of z_0, the point at which the periodic wave function is cut by the surface. In the band gaps, surface states appear (discrete because the model is one-dimensional), and it will be noted that their energy values depend on z_0, and that at certain values they merge into bulk bands.

The Band-Edge Method†

The band-edge method is an extension of the LCAO approximation (Section 6.3) to ionic crystals. Instead of using the same atomic orbital for each atom in the one-dimensional crystal, an s orbital is used for atom M and a p_z orbital for atom X, as shown in Fig. 6.9. As in Section 6.3, the end atom is assigned a Coulomb integral different from the bulk. The s orbitals on the M sites and the p orbitals on the X sites give resonance integrals which alternate in sign along the chain. The method differs from the usual LCAO approach by using conduction and valence band-edge energies as a starting point in computations. These energies are accurately known from a variety of theoretical and experimental studies on MX-type crystals, and they are taken as inputs to the problem of ionic surface

† Ref [23].

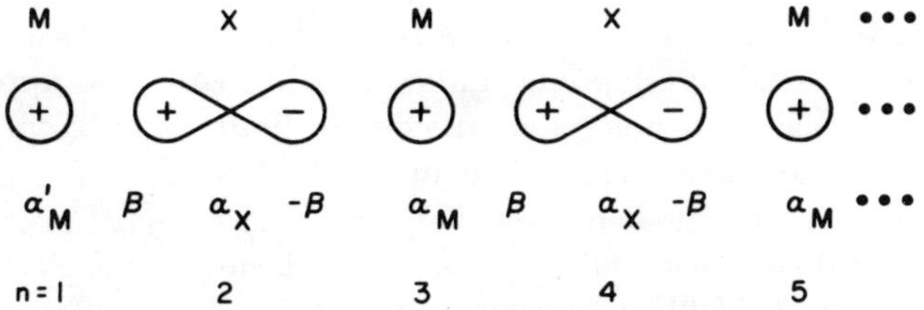

Fig. 6.9. Essential aspects of a one-dimensional ionic semi-infinite chain showing M and X atoms, their orbital wave function symmetry types, Coulomb and resonance integrals, and location with respect to the chain. [After Levine [23]. By permission of the American Institute of Physics.]

states. Such energies and the wave functions associated with them are solutions to the appropriate Schrödinger wave equation. Using this approach, the location of the surface states within the band gap is more accurate. M-like states appear below the conduction band when the one-dimensional lattice is terminated at an M atom, and X-like states appear similarly above the valence band when the lattice terminates at an X atom. It is concluded that the results complement and support the Madelung potential and sinusoidal potential methods.

In a subsequent three-dimensional calculation for the zinc blende (100) surface using an s orbital for the M-site and three p orbitals (p_x, p_y, p_z) for each X site, Levine and Freeman [27] illustrated the importance of surface structures on the energy results obtained. When an undistorted lattice was assumed, no surface bands were found. If all the M atoms are rotated up and all the X atoms rotated down from the ideal surface plane with the bond lengths unaltered, surface bands are found. Since there is strong experimental evidence for the existence of such surface bands, it is concluded that the surface is probably distorted.

Another modified LCAO method to which the band-edge method bears some resemblance is the full-zone **kp** method [28], which has been applied to surface problems [29–31]. Since its surface applications are chiefly to semiconductors of the silicon and germanium type, which are not of paramount interest in studies of chemisorption and catalysis, we shall not describe the method here.

6.5 Conclusion

We conclude this chapter with a brief mention of the problems encountered with metallic systems. In these systems, there is generally broad overlapping of bands, which accounts for their good electrical properties. This prevalent band overlapping makes less probable the appearance of surface bands, that is, bands lying outside the limits of bulk energy bands. It also makes the theoretical and experimental studies of surface states more difficult. The methods of this chapter run into severe problems when applied to metal surfaces, and the semiclassical methods of Section 3.3 are quite commonly used [32]. Of course, the LCAO technique for a monatomic crystal (see 6.2) may be considered an approximation to metal systems [33]. But the popular tight-binding approximation in the LCAO technique, which neglects the overlap integrals of atomic orbitals, is a poor one for metals with their strongly delocalized electrons. Systems other than metals may, under certain conditions, assume metallic-like properties. As an example, we refer to the theoretical case mentioned in Section 6.4, where surface anionic and cationic bands approach each other in the bulk energy band gap until they overlap, creating the possibility of metallic-like surface conductivity. Further discussion of metal surfaces will be found in Chapter 8 in conjunction with chemisorption studies.

REFERENCES

1. I. Tamm, *Phys. Z. Sowjet*, **1,** 733 (1932).
2. R. de L Kronig and W. G. Penney, *Proc. Roy. Soc. London* **A130,** 499 (1931).
3. R. H. Fowler, *Proc. Roy Soc. London* **A141,** 56 (1933).
4. E. T. Goodwin, *Proc. Cambridge Phil. Soc.* **35,** 205 (1939).
5. E. T. Goodwin, *Proc. Cambridge Phil. Soc.* **35,** 221 (1939).
6. E. T. Goodwin, *Proc. Cambridge Phil. Soc.* **35,** 232 (1939).
7. W. Shockley, *Phys. Rev.* **56,** 317 (1939).
8. T. B. Grimley and B. W. Holland, *Proc. Phys. Soc. London* **78,** 217 (1961).
9. J. D. Levine, *Phys. Rev.* **171,** 701 (1968).
10. T. B. Grimley, *Advan. Catal.* **12,** 1 (1960).
11. J. Koutecky, *Phys. Rev.* **108,** 13 (1957).
12. D. Kalkenstein and P. Soven, *Surface Sci.* **26,** 85 (1971).
13. F. Bloch, *Z. Phys.* **52,** 555 (1928).
14. M. Scherer and P. Phariseau, *Z. Phys.* **233,** 368 (1970).
15. C. A. Coulson, *Proc. Roy. Soc.* **A67,** 608 (1954).
16. M. Hoerni, *J. Chem. Phys.* **34,** 508 (1961).
17. P. Mark, *Catal. Rev.* **1,** 165 (1967).
18. T. B. Grimley, *J. Phys. Chem. Solids* **14,** 227 (1960).
19. J. Horiuti and T. Toya, "Solid State Surface Science" (M. Green, ed.), Vol. 1, p. 6. Dekker, New York, 1969.
20. J. D. Levine and P. Mark, *Phys. Rev.* **182,** 926 (1969).
21. J. Koutecky, *Advan. Chem. Phys.* **9,** 85 (1965).
22. S. G. Davison and S. D. Levine, *Solid State Phys.* **25,** 1 (1970).
23. J. D. Levine and S. G. Davison, *Phys. Rev.* **174,** 911 (1968).

24. G. C. Benson, P. T. Freeman, and E. Dempsey, *J. Chem. Phys.* **39,** 302 (1963).
25. D. A. Wright, "Semiconductors," p. 55. Methuen, London, and Wiley, New York, 1950.
26. N. F. Mott and R. W. Gurney, "Electronic Processes in Ionic Crystals," p. 86. Oxford Univ. Press, London and New York, 1950.
27. J. D. Levine and S. Freeman, *Phys. Rev.* (in press).
28. M. Cardona and F. H. Pollak, *Phys. Rev.* **142,** 530 (1966).
29. I. Alstrup, *Surface Sci.* **20,** 335 (1970).
30. R. O. Jones, *Phys. Rev. Lett.* **20,** 992 (1968).
31. I. Bartos, *Surface Sci.* **15,** 94 (1969).
32. N. D. Lang and W. Kahn, *Phys. Rev. B* **3,** 1215 (1971).
33. G. Allan and P. Langlart, *Surface Sci.* **30,** 641 (1972).

VII

Quantum Theory of the Chemisorptive Bond for General Lattices

Having described free-surface states for idealized monatomic systems in the previous chapter, we turn to the surface states created by chemisorption on such surfaces. Many of the phenomena of chemisorption are revealed by these systems, and phenomena not yet observed are predicted. We shall find that the simple one-dimensional model described in Section 7.1 gives a remarkably faithful representation of multidimensional models. The general conditions for localized and nonlocalized bonding states of a single adsorbed atom are derived and extended in Section 7.2 to three-dimensional models. In Section 7.3, the effects of interactions between adsorbed atoms are considered, including an explanation for the observed falling heats of chemisorption with increasing coverage. A brief discussion of the types of surface bonds follows in Section 7.4. Although it is conventional to assume that chemisorption always involves localization of electrons, there is little sound evidence for either localization or nonlocalization of chemisorptive bonds. If adsorption is localized, the bond may be covalent or ionic. The values of interaction parameters that govern the type of localized adsorption bonds are set forth from theoretical considerations. Further, we discuss the nonlocalized or metallic-like bond in which all the electrons in the system, adsorbed atom plus adsorbent, may contribute to the surface bond. There follows in Section 7.5 a qualitative

Fig. 7.1. One-dimensional model for chemisorption

discussion of the heats of chemisorption for both localized and nonlocalized bonding. The chapter closes with a short discussion of the use of finite models with a small number of adsorbent atoms for chemisorption calculations. It is shown that these models provide many of the same conclusions obtained from semi-infinite models.

7.1 One-Dimensional Lattices†

We modify the one-dimensional free-surface model of Fig. 6.5 by adding a foreign atom at one end, as illustrated in Fig. 7.1. The problem we wish to solve is the determination of energy states for the interaction of a hydrogenlike atom with a crystal whose electronic structure consists of a single band of states. Again, we make the tight binding approximation, which neglects overlap integrals between atomic orbitals and includes the resonance integrals only between nearest neighbors (see Section 6.3). The LCAO wave function, $\psi = \sum_r C_r \phi_r$, is the same as before except that it includes the atomic orbital ϕ_λ for the chemisorbed atom, and the one-electron Hamiltonian operator now includes the entire system, chain plus foreign atom. Thus the energy matrix obtained by the variational procedure has all of the elements—Coulomb and resonance integrals—of the free-surface model and, in addition, the Coulomb and resonance integrals for the foreign atom

$$H_{\lambda,\lambda} = \alpha'', \qquad H_{0,\lambda} = \beta'. \tag{1}$$

The foreign atom is characterized by a different Coulomb integral α'' from that of the crystal atoms ($\alpha = H_{rr}$ when $r \neq 0$, and $\alpha' = H_{00}$ when $r = 0$). The resonance integral between the foreign atom and the end crystal atom is denoted by $\beta' = H_{0,\lambda}$ to distinguish it from $\beta = H_{r,r\pm1}$, $r = 0, 1, 2, \ldots N$, for crystal atoms. The energy equation is identical in form to that for the free-surface model, Eqs. (11) and (14), Section 6.3, but we now have two boundary conditions instead of one,

$$(E - \alpha')C_0 = \beta C_1 + \beta' C_\lambda, \tag{2}$$

$$(E - \alpha'')C_\lambda = \beta' C_0. \tag{3}$$

For our purposes, it is not necessary to describe the method of solving

† Refs. [1–3].

the energy equation with the boundary conditions Eqs. (2) and (3). A few details of the solution for a free surface are given in Section 6.3, and the pattern here is similar. After introducing the dimensionless quantities

$$z = (\alpha - \alpha')/\beta, \qquad z' = (\alpha - \alpha'')/\beta, \qquad \eta = \beta'/\beta, \qquad E' = (E - \alpha)/2\beta,$$
$$-1 < E' < 1, \tag{4}$$

we obtain

$$(z + \cos\theta + \sin\theta \cot N\theta)(z' + 2\cos\theta) = \eta^2, \tag{5}$$

which may be compared to the similar equation for the free-surface model, Eq. (18), Section 6.3. As before, we denote localized states by $\mathcal{P}$ or $\mathfrak{N}$, according to whether E' is positive or negative. These are localized states associated with the foreign atom and the crystal atoms near the surface. The occurrence of these states is governed by the values of three parameters, z, z', and η, which define the nature of the interaction between the foreign atom and the crystal. An analysis of Eq. (5) with N large shows that localized states exist for given η in regions of the zz' plane defined by the two hyperbolas

$$(z \pm 1)(z' \pm 2) = \eta^2. \tag{6}$$

In Fig. 7.2, we plot these two hyperbolas for $\eta^2 = 1$. Localized states occur,

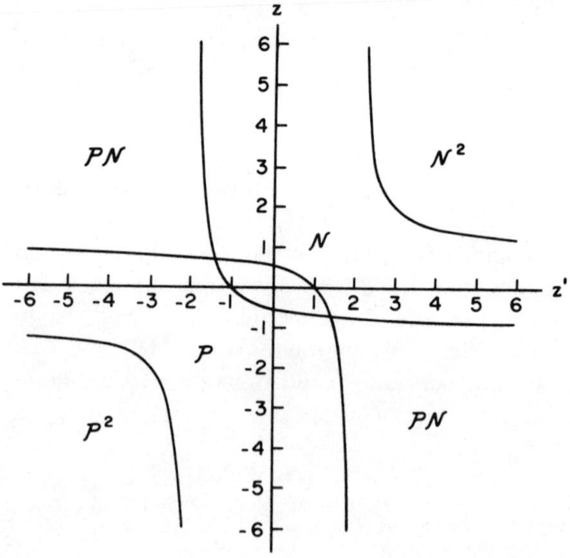

Fig. 7.2. The occurrence of localized states for $\eta^2 = 1$ [1].

as indicated, in the following regions:

1. region $\mathcal{P}^2$—two localized states;
2. region $\mathcal{N}^2$—two localized states;
3. region $\mathcal{P}$—one localized state;
4. region $\mathcal{N}$—one localized state;
5. two regions $\mathcal{PN}$—one localized $\mathcal{P}$ state and one localized $\mathcal{N}$ state.

Thus, the system may have zero, one, or two localized states. The area of the region where no localized states exist decreases as η increases. Only if $\eta^2 < 2$ does such a "forbidden" region exist.

We see that the maximum number of localized states that can be formed is two. But this result is governed by our assumption that only one orbital on the foreign atom and one band of crystal orbitals are interacting, and that perturbations by the foreign atom are limited to the first atom of the crystal. If the Coulomb integrals are modified to include perturbations of the first and second crystal atoms, a maximum of three localized states may exist. In general, the maximum number of localized states is equal to the sum of the number of orbitals on the foreign atom, the number of crystal atoms perturbed, and the number of bands of crystal states. From the simple one-dimensional model, we find many features of the general chemisorption problem.

If the atomic orbitals of the crystal used in the LCAO approximation are all s orbitals, then β, the resonance integral, is negative, and the $\mathcal{P}$ states will lie below the crystal band and the $\mathcal{N}$ states above it, as explained in Section 6.3. For this reason, the $\mathcal{P}$ levels are called bonding states and the $\mathcal{N}$ levels antibonding states, in analogy with the terminology for the states of diatomic molecules. If the atomic orbitals of the crystal are p_z orbitals with the z axis along the chain, β is positive and, conversely in this case, the $\mathcal{N}$ states are bonding and the $\mathcal{P}$ states antibonding. In Table 7.1, the bonding states for several types of orbitals are summarized. We take the z axis along the chain and use the symmetry symbols σ and π, as in the description of the bonding orbitals of diatomic molecules. The σ states are symmetrical about the chain axis and the π states have a nodal plane passing through the same axis. The classifications in the table hold for a three-dimensional crystal if the chemisorbed atom is located directly above a surface atom of the crystal; otherwise the problem is more complicated.

Variations of the simple one-dimensional model presented above have been studied. For example, Davison and Koutecky [4] (using the LCAO–MO tight binding approximation) analyzed the electron states associated with the surfaces of one-dimensional mixed crystals having alternately

TABLE 7.1

Bonding States for a Foreign Atom and a Chain[a]

Foreign atom orbital	Chain band	Bonding state
s	s, d_{z^2}	$\mathcal{P}\sigma$
s	p_z	$\mathcal{N}\sigma$
p_x	p_x, d_{zx}	$\mathcal{P}\pi$
p_y	p_y, d_{yz}	$\mathcal{P}\pi$
p_z	p_z	$\mathcal{N}\sigma$
p_z	d_{z^2}	$\mathcal{P}\sigma$

[a] Grimley [1].

atoms of A and atoms of B. In this case, the energy levels lie in two bands separated by a gap whose width is characteristic of the mixed crystal; and surface and chemisorption states may have energies lying either outside the two main bands or in the gap between them.

7.2 Three-Dimensional Lattices†

Differences between the results of calculations for one-dimensional and multidimensional crystals must be anticipated. For example, surface states for the three-dimensional crystal are no longer discrete levels, but bands. Further, full coverage of the surface of the one-dimensional crystal requires only one foreign atom, whereas N foreign atoms, one for each crystal surface atom, are required for the three-dimensional crystal. These spatial differences add complexity to the multidimensional case. The solution of the energy equation is far more difficult and it cannot be solved directly. But a Green's function solution can be obtained by the methods developed by Baldock [7] and by Koster and Slater [8], still using the tight binding approximation. As in the one-dimensional case, the results depend on the parameter z, which governs the existence of surface states of the crystal alone, and other parameters. Yet in spite of added complexities, the similarities of the one-dimensional and three-dimensional results are striking, which, of course, justifies our initial consideration of the simpler model.

Let us assume first that the three-dimensional crystal before chemi-

† Refs. [1–6].

sorption has no surface states ($|z| < 1$). Just as in the one-dimensional case, a chemisorbed foreign atom may create localized surface states lying outside the crystal band. The wave functions for these states are appreciable on the foreign atom and on the surface atoms of the crystal upon which it is adsorbed, but fall to zero in all directions beyond the range of perturbation. A maximum of two localized states may be formed, and depending on the values of the interaction parameters, there may be only one or even none. In the last case, all the states of the system, foreign atom plus crystal, are nonlocalized, their wave functions periodic in the crystal, and their energies all in the crystal band. Localized states are called $\mathcal{P}$ or $\mathcal{N}$, depending on whether they have positive or negative values of E'. If the resonance integral β is negative, $\mathcal{P}$ states have energies lying below the crystal band, and $\mathcal{N}$ states lying above it. The converse is true if β is positive. Grimley [2] has shown that regions of zero, one, and two localized states are bounded by hyperbolas, as in the one-dimensional case (Fig. 7.2), but that the area of the region of nonlocalized states, where localized states are forbidden, is considerably greater in the three-dimensional model. From Section 7.1, we see that the essential features of the one-dimensional model are retained.

We assume next that the crystal before chemisorption has surface states lying outside of the normal crystal band of energy states ($|z| > 1$). The center of the band of surface states is displaced from the center of the normal band of crystal states, and when z is large, the surface band is completely separated from the main band. As z approaches infinity, the amplitudes of the wave functions of the normal crystal states on the surface atoms of the crystal tend to zero. Only the surface states have wave functions of finite amplitude on the crystal atoms of the surface. Consequently, chemisorbed atoms essentially ignore the bulk crystal and its energy band. As far as chemisorption is concerned, the crystal acts as though it were a two-dimensional array of atoms with the surface band playing the role of the main band. If there are localized chemisorption states, they will separate out of the surface crystal band. A priori, it might be considered that for certain values of the parameters a chemisorbed atom might interact with all of the surface atoms of the crystal, forming nonlocalized states whose energies all lie in the crystal surface band, and for other values, localized states might form, representing interactions of the chemisorbed atom with one or a small number of surface atoms. That is what would be expected on the basis of results up to this point, which all indicate a region of nonlocalized states, where localized states are forbidden. But theoretical calculations [2] do not bear this out. In the present case, the hyperbolas in the zz' plane do not intersect for any values of the parameters, which means simply that localized states will always occur, that the forbidden region has shrunk to zero.

We note that so far the discussion of the chemisorption of a foreign atom is restricted to the use of a single orbital for each crystal atom. But additional bands may be formed, one for each additional orbital. These bands may interact and, especially in the case of metals, overlapping occurs. The use of more than one orbital on the foreign atom complicates the picture even further. Localized states must be outside of all bands. Although the multiple band picture does not present any formal difficulties, the resulting equations are so complex that useful information is not extracted easily from them. There is a pressing need to extend calculations in this direction.

At the beginning of this section, we mentioned that the strict generalization of the one-dimensional model of chemisorption (Fig. 7.1) leads to a three-dimensional crystal with its surface completely covered by adsorbed atoms. We wish to discuss the saturated model in this section, for it has a relatively simple solution and leads to an important new feature, aside from adsorbate–adsorbate interactions, which will be taken up in the next section. The problem is treated by the simple LCAO approximation, as in the one-dimensional case, by changing the Coulomb integral from its value α in the interior of the crystal to α' on the crystal atoms at the surface and to α'' on the atoms of the adsorbed layer. Also, the resonance integral between nearest neighbors in the interior is β; between a surface crystal atom and an adsorbed atom β'; and between nearest neighbors in the adsorbed layer, β''. The solution of the energy equation for a simple cubic lattice follows in a manner similar to that for the one-dimensional solution. The interaction parameters z, z', and η are defined as in Section 7.1 for the one-dimensional case; and a new parameter, $\eta' = \beta''/\beta$, is added. We shall consider first the case where no surface states exist before chemisorption ($|z| < 1$), all states of the crystal surface atoms lying within the normal crystal band. Furthermore, if $\eta' = 1$, that is, the resonance integral between nearest neighbors in the adsorbed layer is the same as in the crystal itself, then the solution is essentially identical with that of the one-dimensional model. The only difference is that the discrete states $\mathcal{P}$ and $\mathfrak{N}$ of the one-dimensional model now become bands of surface states associated with the adsorbed layer and the crystal surface. No more than two bands may be formed, and each band contains N^2 levels, where N is the number of atoms in the adsorbed layer. Depending on the values of z and z', these bands may or may not overlap the normal crystal bands. All of this is reminiscent of the one-dimensional model, and in fact, Fig. 7.2 describes the situation for the three-dimensional case with $\eta^2 = 1$ and $\eta' = 1$. But when η' does not equal unity, a new feature arises with an energy condition different from that of the one-dimensional model. We now find that in place of z' a new set of values exists whose limits of variation from the original value of z' are given by $z' + 4(1 - \eta')$ and $z' - 4(1 - \eta')$. Referring to Fig. 7.2, we see that these changes in values of z' might force us out of the $\mathcal{P}$ region into the nonlocalized region. If this happens, we do not get a complete band of $\mathcal{P}$ states containing N^2 levels. The missing states fall into the normal band of crystal states and so are no longer localized. If the variation in z' takes us through the nonlocalized region and into the $\mathfrak{N}$ region, then there will be an incomplete $\mathcal{P}$ band and an incomplete $\mathfrak{N}$ band with less than N^2 levels in each.

When the energy states of the surface crystal atoms lie outside of the normal crystal band ($|z| > 1$), a complete band of N^2 levels of one sort or another ($\mathfrak{N}$ or $\mathcal{P}$) will always exist in the surface bands. For example, there may be an incomplete $\mathfrak{N}$ band and a complete $\mathcal{P}$ band; other possibilities can be found in Fig. 7.2.

7.3 Interaction between Chemisorbed Species†

We observed in the previous section that interesting and novel phenomena occur when the surface is fully covered by chemisorbed atoms. To analyze the situation in greater detail, we return to a simpler system with only two chemisorbed atoms on a chain of crystal atoms. Unlike the strictly one-dimensional model of Section 7.1 with chemisorption only on the end atom, the present model allows adsorption on every atom of the chain, as if it were a row of atoms lifted from the surface of a three-dimensional crystal and numbered $-N, -(N-1), \ldots, -2, -1, 0, 1, 2, \ldots, (N-1), N$. We assume that the crystal alone has no surface states. If the adsorbed atoms are infinitely far apart, any localized levels which are formed will be doubly degenerate. At finite separation, interactions involving both chemisorbed atoms and the crystal surface will occur. The wave functions for the system are a linear combination of atomic orbitals of the crystal and of the two chemisorbed atoms. These wave functions of the system are either even or odd with respect to the center of symmetry at the chain atom numbered 0, and each state, even or odd, has a solution similar to that of the strictly one-dimensional model in Section 7.1. The even and odd solutions may have localized states associated with the crystal surface and both foreign atoms. As before, we shall refer to these

† Refs. [1, 3, 5, 6].

localized states as $\mathcal{P}$ states (odd or even) and $\mathfrak{N}$ states (odd or even). The states are bonding if β, the resonance integral, is negative, and will separate below the normal crystal band.

Certain general conclusions may be drawn from the energy conditions set by the even and odd solutions. If there is a single $\mathcal{P}$ state for one atom on the surface (see Fig. 7.2), there will always be an even $\mathcal{P}$ state when two foreign atoms are present; and if there are two $\mathcal{P}$ states for one foreign atom on the surface, there will be two even $\mathcal{P}$ states when two foreign atoms are present. The energies of the even $\mathcal{P}$ states fall below those of the corresponding $\mathcal{P}$ states for a single atom on the surface, and fall lower as the distance of separation of the chemisorbed atoms decreases. Conversely, the odd $\mathcal{P}$ states for two atoms on the surface lie above the corresponding $\mathcal{P}$ states for a single atom, and their energies rise as the distance of separation decreases. An odd $\mathcal{P}$ state may rise sufficiently to enter the bottom of the normal crystal band and thus become a nonlocalized state. It may even pass through the normal crystal band and appear as an $\mathfrak{N}$ state above the band. Similar, but reverse, arguments can be made for $\mathfrak{N}$ states, which always lie above the crystal band when $\beta < 0$.

To clarify the picture, we describe qualitatively what happens as two chemisorbed atoms approach each other. Assume first that a single chemisorbed atom forms a $\mathcal{P}$ state with two localized electrons in it. Now, a second atom is chemisorbed at a point remote from the first atom. As the two atoms approach each other, the identical individual $\mathcal{P}$ states split into an even and an odd localized state associated with both atoms. The even localized state lies below the original degenerate level and the odd state lies above it. Both localized levels contain two electrons. Bringing the two chemisorbed atoms still closer together pushes the odd $\mathcal{P}$ state higher and higher. Eventually, it may become a nonlocalized state in the normal crystal band. Note that there is a mechanism here for falling heats of chemisorption with increasing coverage.

We point out that in the solution of this model of chemisorption, the resonance integral between the two chemisorbed atoms was assumed to be zero at all separations. Therefore, there is no direct interaction between them. The important interaction occurs *through the crystal*. If the localized level for a single adsorbed atom does not lie too far below the band of crystal states, this interaction is large. The spread of the wave function along the chain parallel to the surface, which measures the amount of interaction, will be greatest when the two atoms are chemisorbed on nearest-neighbor sites. The wave function of the odd $\mathcal{P}$ state is less well localized than that of the even $\mathcal{P}$ state along the chain, for the former is closer to the bottom edge of the crystal band. Such indirect interactions between two chemisorbed atoms will take place at distances where direct interaction would be negligible.

Quantum-mechanical calculations have been made recently [9] for the interaction of two closed-shell atoms on a flat surface constraining adsorbed atoms to two-dimensional movement, and on a periodic surface of deep potential wells. It was found that the flat model gives very weak adsorption at ordinary temperatures.

7.4 Types of Surface Bonds†

In molecular orbital theory, the electronic ground state of the system, crystal plus foreign atom, is obtained by successive filling of the levels with two electrons in each, starting with the lowest level. If all filled levels lie in the crystal band, there will be no localized surface bonds. But if a filled level lies outside of the crystal band, then a localized surface bond will exist. This bond may vary from purely covalent to ionic in character. Following Grimley [1], we shall describe qualitatively in this section how the interaction parameters determine the character of the chemisorption bond. For simplicity, the one-dimensional model of Section 7.1 will be used. It contains all of the important features of the three-dimensional model and is far easier to handle.

The Covalent Bond

When it is equally probable to find an electron on the chemisorbed atom as on atoms of the crystal and when the corresponding energy level contains two electrons, we have a purely covalent surface bond. The number of crystal atoms involved may vary from one—the one directly below the adsorbed atom—to many. In the former case, the bond is called a two-center covalent bond, and in the latter case, a many-center covalent bond. In either case, the nature of the bond is described by the quantity R, the *charge order* of the state on the adsorbed atom,

$$R = |C_\lambda|^2 \Big/ \left[|C_\lambda|^2 + \sum_{r=0}^{N} |C_r|^2 \right] \tag{7}$$

where the constants C are the coefficients of the wave functions in the linear combination of atomic orbitals mentioned in Section 7.1. The constant C_λ is the coefficient of the wave function of the foreign atom and the C_r are the coefficients of the wave functions of the crystal atoms. For a

† Ref. [1].

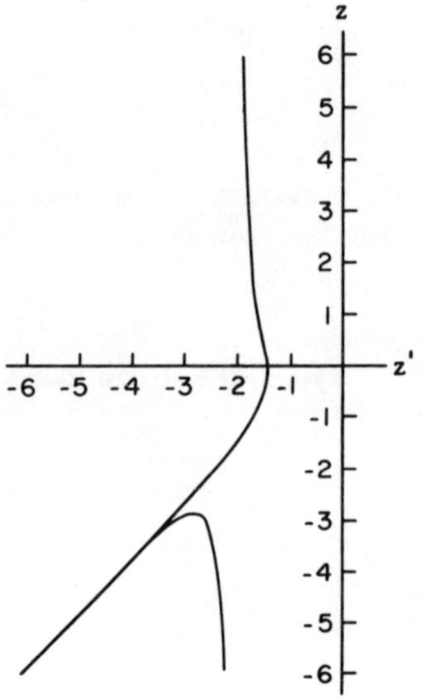

Fig. 7.3. The occurrence of covalent $\mathcal{P}$ states for $\eta^2 = 1$ [1].

purely covalent bond, $R = \frac{1}{2}$, which signifies that the electrons of the localized surface bond spend one half their time on the foreign atom and one half on the crystal atoms. In order to consider a specific case, we take β, the resonance integral, to be negative, for which the $\mathcal{P}$ states are bonding and lie below the crystal band of states. Pairs of values of z and z' (see Eq. (4) and Fig. 7.2) may be calculated with $\eta^2 = 1$ and $R = \frac{1}{2}$, representing covalent $\mathcal{P}$ states. They lie on the curves in Fig. 7.3. There are two branches, each of which approaches the line $z' = z$ and the line $z' = -2$ asymptotically. Comparing positions on Fig. 7.2 and on the curves of Fig. 7.3, we can find the number of $\mathcal{P}$ surface states that exist. Further, knowing z, z', and η, it is possible to calculate the energy levels and the wave function coefficients C. Knowing the wave function coefficients, it is

TABLE 7.2

Wave Function Coefficients for a Covalent $\mathcal{P}$ State[a]

r	λ	0	1	2	3	4	5	6
C_r	2.350	1.000	0.954	0.863	0.781	0.707	0.604	0.579

[a] From Grimley 1]; $z = 1.245$, $z' = -1.585$, $\eta = 1$.

TABLE 7.3

Wave Function Coefficients for a Covalent $\mathcal{P}$ State[a]

r	λ	0	1	2	3	4
C_r	1.026	1.000	0.223	0.050	0.011	0.002

[a] From Grimley [1]; $z = -3.46$, $z' = -3.72$, $\eta = 1$.

possible to show how the interaction of the chemisorbed atom with crystal atoms falls off with distance. Grimley [1] gives an example with $z = 1.245$, $z' = 1.585$, and $\eta = 1$. This is a location high on the upper branch in the $\mathcal{PM}$ region of Fig. 7.2. One $\mathcal{P}$ state exists, and Grimley shows that the corresponding energy level is just below the bottom edge of the crystal band, which means that the interaction of the foreign atom with the crystal should fall off slowly with distance into the crystal. Table 7.2 shows clearly the slow dampening of the wave function in the crystal for this location on the curve $R = \frac{1}{2}$ in Fig. 7.3. The many-center character of this surface bond is evident. Moving down the curve in Fig. 7.3 to $z = -3.46$, $z' = -3.72$, and $\eta = 1$, we enter the $\mathcal{P}^2$ region of Fig. 7.2. Considering the lower $\mathcal{P}$ state, which is now far below the bottom edge of the crystal band, we find the interaction more restricted. In fact, the surface bond, as shown in Table 7.3, where the wave functions fall off rapidly with distance into the crystal, is essentially a two-center covalent bond.

On the lower branch of the curves in Fig. 7.3, two $\mathcal{P}$ states always exist, for the branch lies completely in the $\mathcal{P}^2$ region of Fig. 7.2. Only one of these states is purely covalent—the one of greatest separation below the crystal band.

Ionic Bonds

If the electrons in a $\mathcal{P}$ state spend more time or less time on the chemisorbed atom than they spend on the crystal atoms, then the surface bond is said to have ionic character. Again, the character of the bond is measured quantitatively by the value of R. The ratio R given by Eq. (7) lies between zero and unity. The localized states are classified as anionic if R is greater than one half and as cationic if R is less than one half. We see that for a bond with anionic character, an electron is concentrated more on the foreign atom, and the reverse is true for a bond with cationic character. On the basis of information in Chapter 3, we note that ionic character and location of the localized states relative to the crystal band of states are of great importance in the theory of chemisorption on semiconductors and insulators—a subject which we shall take up again in Chapter 9.

METALLIC BONDS

If the interaction parameters z and z' fall in the forbidden region in Fig. 7.2, no localized surface bonds are formed between the chemisorbed atom and the crystal atoms. All energy states of the system lie in the normal crystal band and all electrons in the system contribute to the surface bond. Strong binding of the foreign atom to the crystal is still possible by virtue of the general lowering of the energies of the electrons caused by the presence of the foreign atom, which alters the boundary condition at the free surfaces. Since the nonlocalized chemisorptive bond is more characteristic of the binding in metals, it is commonly referred to as metallic-like, and we shall discuss it in more detail in the next chapter.

7.5 The Heat of Chemisorption

In the following discussion, we shall use the terms heat of chemisorption and energy of chemisorption synonymously. When the chemisorptive bond is localized, the definition of the heat of chemisorption is straightforward. It is the energy value of the localized level relative to zero for the crystal and foreign atom infinitely separated. But when the bond is nonlocalized and all the electrons in the system, foreign atom plus crystal atoms, participate in the bonding, the definition of the heat of chemisorption and the procedure for calculating it are more complex.

To get an idea of the magnitude of the energy change associated with nonlocalized chemisorption, we refer again to the one-dimensional model of Section 7.1. It is assumed that there are N electrons, one for each atom in the chain, which means that the band is half filled. With the foreign atom removed to infinity, the total electronic energy of the crystal chain, E_N, is calculated by the procedure of Baldock [10]. The calculation assumes that no free surface states are occupied. We now bring the foreign atom up to the free end of the chain and make a similar calculation of E_{N+1}, the total electronic energy for the system of N crystal atoms and one chemisorbed foreign atom. For simplicity, it is assumed in making this calculation that $\beta' = \beta$ where β' is the resonance integral between the foreign atom and the end crystal atom, and β that between all other crystal atoms; and that $\alpha' = \alpha$, where α' is the Coulomb integral on the end crystal atom and α that on all the other crystal atoms. Now, if ϵ is the energy of the valence electron on the isolated foreign atom, we find for ΔE, the change in total electronic energy when the foreign atom and the chain interact,

$$\Delta E = E_{N+1} - E_N - \epsilon. \tag{8}$$

Without going into mathematical detail, we find, on substitution of expressions for E_{N+1} and E_N in Eq. (8), that

$$\Delta E = [\alpha + (4\beta/\pi) - \epsilon] + \Delta E_s \tag{9}$$

where ΔE_s is the difference between the surface terms in the expressions for E_N and E_{N+1} and gives the change in surface energy caused by the presence of the chemisorbed atom. The term in brackets is the difference between the mean electron energy in the crystal and the energy of the valence electron in the isolated foreign atom. This energy is associated with the delocalization of the valence electron on the adsorbed atom. It turns out that the two terms in Eq. (9) have opposite signs and tend to cancel each other partially. Numerical values of ΔE cannot be calculated yet for particular cases because we have no knowledge of the numerical values of the interaction parameters. However, qualitative deductions for specific metal systems have been made and will be discussed in the next chapter.

7.6 Small Finite Models

We have seen that workers using semi-infinite models, both one-dimensional and multidimensional, have come to the important theoretical conclusion that localized electron states appear for certain values of the interaction parameters. Other workers have used small finite models [11, 12]. For example, Blyholder and Coulson [11] have used a one-dimensional model consisting of only six atoms. When the end atom of the chain is given a different Coulomb integral, they find that the essential features of the semi-infinite model are duplicated in a simple LCAO calculation. Of course, a band of levels is not found with this small number of atoms; nevertheless, a localized orbital appears in the neighborhood of $|z| = 1$. Thus, the small finite model shows the separation of a localized state at approximately the same values of z as the semi-infinite model. Furthermore, it was shown that the rate of decrease of the LCAO wave function coefficients is independent of chain length for chains of 6, 8, 10, and 12 atoms in the region of a strongly localized state ($z = 2$). A more elaborate LCAO calculation gives essentially the same results, thus fostering confidence in the simple, tight binding LCAO technique. Finally, they showed that although the choice of adsorption site on a surface depends strongly on the properties of the appropriate localized state, it is also important to include the other electrons in order to avoid error. The small finite model is far simpler to handle in this case, as Blyholder and Coulson demonstrated by using a 6 × 6 square array of atoms with a single atom adsorbed on it at various locations.

REFERENCES

1. T. B. Grimley, *Advan. Catal.* **12,** 1 (1960).
2. T. B. Grimley, *Proc. Phys. Soc. London* **72,** 103 (1958).
3. J. Koutecky, *Advan. Chem. Phys.* **9,** 85 (1965).
4. S. G. Davison and J. Koutecky, *Proc. Phys. Soc. London* **89,** 237 (1966).
5. T. B. Grimley, *J. Phys. Chem. Solids* **14,** 227 (1960).
6. J. Koutecky, *Trans. Faraday Soc.* **54,** 1038 (1958).
7. G. R. Baldock, *Proc. Cambridge Phil. Soc.* **48,** 457 (1952).
8. G. F. Koster and J. C. Slater, *Phys. Rev.* **95,** 1167 (1954).
9. A. Bagchi, *Phys. Rev.* A, **3,** 1153 (1971).
10. G. R. Baldock, *Proc. Phys. Soc.* **A66,** 1 (1953).
11. G. Blyholder and C. A. Coulson, *Trans. Faraday Soc.* **63,** 1982 (1967).
12. H. H. Dunken and Ch. Optiz, *Z. Phys. Chem. Neue Folge* **60,** 25 (1968).

VIII

Quantum Theory of the Chemisorptive Bond for Metals

The hypothetical lattice model of the last chapter brings out many of the characteristics of chemisorption on solid surfaces. But there is, of course, a limit to its usefulness in explaining real systems. Each of the four types of solids—molecular crystals, covalent crystals, ionic crystals, and metallic conductors—has its peculiarities, and within each of these broad types, many differences exist in composition and kinds of imperfections. No general theory encompasses all situations. To progress, we must make allowances for the structural peculiarities of real solids. In this chapter, we shall be concerned with metals. The approach remains basic and only occasionally will reach that degree of detail necessary for the consideration of particular metals. We must be content for the present if such basic theories can make general distinctions in the nature of chemisorption on the various classes of solids.

The description of electron levels in solids in terms of allowed and forbidden zones is applied to all periodic systems, metallic or otherwise. But a metal, or more specifically, a conductor, has a half-filled band, overlapping of completely full and completely empty bands, or more complex combinations, which account for electrical conductivity. Each atom of a metal is regarded as ionized in a sea of electrons. Unlike heteropolar crystals, where the individual units are more or less distinct positive and negative ions held together by electrostatic forces, the binding in metals is usually considered to be the result of an approximately uniform electron cloud. The characteristic electronic structure of metals profoundly influences the nature of chemisorption on their surfaces. The use of only one atomic orbital per metal atom, as in the previous chapter, precludes all effects of orbital overlapping so prevalent in metals. Transition metals bring up the additional question of the effect of empty d orbitals on chemisorption (Chapter IV).

In Section 8.1, we summarize briefly those theories of metals which have been used in conjunction with studies of chemisorption, some of which have been discussed in earlier chapters. In Section 8.2, we discuss the theoretical studies of electronic states created by chemisorption on metal surfaces. As mentioned many times, the two basic approaches involve (1) the assumption of a localized bond and (2) treatment of the entire crystal plus foreign atom. In the latter approach, we have seen that both localized and nonlocalized states may be predicted. The localized approach has been treated in earlier chapters, but in this chapter we shall pursue further the concept of a localized surface compound, ionic and covalent, from a more fundamental point of view. In the last section, we shall discuss the electron density in a metal near a chemisorbed atom or molecule. Theoretical calculations suggest that electronic disturbances by a foreign atom near a metal surface are long-range, oscillatory, and nonisotropic.

8.1 Theories of Metals in Relation to Chemisorption

Every theory of chemisorption on metals envisions a picture of the substrate. These pictures range from oversimplified and qualitative to detailed and quantitative. One of the simplest theories of metals, the free-electron theory (Section 3.1), in which a metal is described as a potential box of mobile valence electrons, leads to a theory of ionic chemisorption. The metal has no structure, its surface is smooth and nonperiodic, and adsorption is governed by the work function of the metal, the ionization potential or electron affinity of the adsorbate, and the mirror-image potential between metal and adsorbate (Section 3.2). Another theory, the valence-bond theory, or rather a simplified empirical version of it (Section 2.2), assumes that d electrons in metals are localized either on atoms or in bonds and leads to qualitative theories of covalent chemisorption using d-orbital vacancies of transition metals (Chapter IV) or "dangling" sp and dsp hybridized orbitals emerging from the surface. In other studies, the theory of chemisorption itself defines a simplified model of the metal. For example, when it is assumed that the foreign atom interacts with only one or a small cluster of surface metal atoms, the model of the metal is automatically limited to the number of metal atoms involved in the chemisorption process, although these few atoms are usually endowed with a metallic character. Chemisorption on small domains has been studied by various quantum-mechanical procedures, including the simple diatomic molecule approach of Higuchi et al. [1] (Section 3.3), the extended Hückel LCAO–MO method [2] (Section 2.3), the three-particle perturbation approach of Jansen [3] for d metals, and the surface compound concept of Grimley [4] (Section 8.2).

Results of broad interest develop when the metal is considered to be a semi-infinite periodic structure with which adsorbate particles interact. A few methods in this category are empirical, giving only a nebulous picture of the metal. For example, the Lennard-Jones [5] potential has been applied to interactions of foreign atoms with a metal. It consists of attractive terms related to the dispersion forces and derived by London using quantum-mechanical methods. According to London's theory, electrons in atoms and molecules are in continuous motion even in their ground states, and thus possess rapidly fluctuating dipole moments. The fleeting dipole moment in one atom perturbs a neighboring one, inducing a moment in it; and the temporary moment in the first atom and the moment induced in the second lead to an attractive force between them. Dipole–dipole interactions lead to attractive potentials which are proportional to the inverse sixth power of the distance between the two atoms. Occasionally, dipole–quadrupole and quadrupole–quadrupole interactions are also included, which vary inversely as the eighth and tenth power of the distance, respectively. Attractive potentials are combined with a repulsive potential, which is taken to vary inversely in the range of the ninth to the fourteenth power of the distance of separation, usually the twelfth power. The total potential between two atoms, including dipole–dipole and dipole–quadrupole attractive potentials, is expressed by

$$u_{\mathrm{d}} = -(C_1/r^6) - (C_2/r^8) + (C_3/r^{12}). \tag{1}$$

Less frequently, an exponential repulsive potential proposed by Born and Mayer [5, 6] is employed; it has the form Be^{-ar}.

The interaction of an adsorbed atom with the metal lattice is determined by pairwise summation of the interactions of the foreign atom with each of the atoms of the metal. The summation may be carried out directly or by integration. In many cases, the nearest pairwise interactions are summed directly and the remainder by integration. When integration is performed , the lattice is tacitly assumed to be a semi-infinite continuum

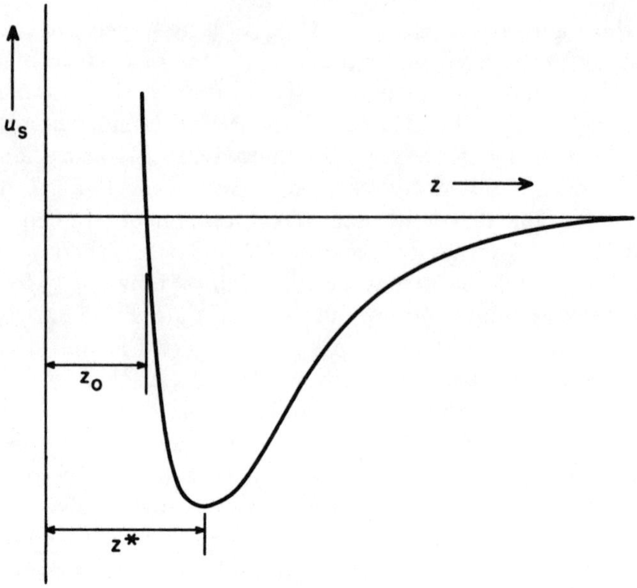

Fig. 8.1. Lennard-Jones 6–12 integrated pair potential for an adsorbed molecule.

with no periodic structure. Using only the dipole–dipole attractive term $-C_1/r^6$ and the repulsive term C_3/r^{12}, integration gives, for the Lennard-Jones potential,

$$u_s(z) = (3\sqrt{3/2})\epsilon_s[(z_0/z)^9 - (z_0/z)^3] \tag{2}$$

where z is the vertical distance of an adsorbed atom from the surface, $u_s(z_0) = 0$, $(\partial u_s/\partial z) = 0$ at $z = z^*$, and $u_s(z^*) = -\epsilon_s$, the potential energy minimum. The potential curve is plotted schematically in Fig. 8.1. The parameters of Eq. (2), ϵ_s and z_0, which are equivalent to the use of the constants C_1 and C_3, may be determined by comparison with measured physical properties, such as virial coefficients and viscosity data. Values of the equilibrium distance z^* of the adsorbed molecule from the surface are often assumed on the basis of some reasonable argument, and values of the parameter C_1 may be determined by theoretical equations developed by London [5, 7], Slater and Kirkwood [5, 8], and Mueller [5, 9]. The dipole–dipole attractive potential $-C_1/r^6$ of Eq. (1) represents weak, or van der Waals, adsorption, and fits more closely the category of physical adsorption than chemisorption. Application to metal surfaces is complicated by the nonlocalized character of the electrons in metals to which the approximations of the Lennard-Jones potential are not strictly amenable. The empirical nature of the method leads only to a nebulous picture of the metal. For this reason, we shall not discuss it further.

Olander [10] has used a similar procedure to describe the interaction of hydrogen atoms with nickel and copper. He replaces the Lennard-Jones potential with a Morse potential, which is more realistic for chemisorptive binding on metals, and is represented by the function

$$\phi(r) = D\{\exp[-(2m/r_e)(r - r_e)] - 2\exp[-(m/r_e)(r - r_e)]\} \quad (3)$$

where $\varphi(r)$ is the potential between two atoms separated by distance r, and r_e, m, and D are force constants. As in the use of the Lennard-Jones potential, the pairwise interactions $\varphi(r)$ are summed over the metal lattice. But in spite of the greater relevancy of the Morse potential to chemisorption and to metals, the method is empirical and the detailed nature of the metal structure in this theory remains nebulous.

Among the more basic theories of metals, we have already discussed the band theory in Chapter VI and its relation to chemisorption in Chapter VII. Our approach in these chapters was general. For the most part, the theory was described in its simplest terms, namely, one hydrogenlike orbital on each metal atom and one similar orbital on the foreign atom. But this model, as we have pointed out before, is not sufficient to describe some of the important characteristics of real metals, such as band overlapping (Section 4.1). Using more than one orbital per metal atom in the LCAO–MO approach, which creates the possibility of band overlapping, brings about great changes in the parameters governing the appearance of localized levels. If broad overlapping of bands occurs, so that a quasi-continuous set of levels extends over an extremely wide energy range, then the appearance of localized levels might be highly improbable. The quantum-mechanical calculations of chemisorption states on models of metals involving more than one atomic orbital per metal atom are extremely difficult. Indeed, most of the band-theoretical conclusions concerning chemisorption on realistic models of metals are merely qualitative or empirical extensions of the results of simpler models involving one atomic orbital per atom. We shall consider these extensions in the following section.

Recently, Johnson [11] proposed an interstitial-electron model for the structure of metals and alloys, and has speculated on its application to chemisorption and catalysis. We shall review it briefly. In this theory, the itinerant electron density is considered to be localized in lattice interstices. Direct comparisons with the band theory are difficult, for in that theory energy–momentum coordinates are used, whereas the interstitial-electron theory is developed in terms of spatial coordinates. Among the first to propose an interstitial-electron model for metals were Lorentz

[12], Fajans [13], and Bent [14]. The novel feature of Johnson's work [11] is the definitive description of the metallic state by an interstitial-electron model which incorporates all the quantum-mechanical aspects of chemical binding, including electron correlation and spin. His work is based on Berlin's [15] analysis of the Hellmann–Feynman theorem, which points out that at equilibrium the classical Coulomb forces exerted by quantum-mechanical electron distribution on the nuclei must balance precisely the forces of nuclear–nuclear repulsion. Extension of the model to metals allows the treatment of metals in a qualitative or quantitative manner on an electrostatic basis.

We summarize the basic postulates of the interstitial-electron model applied to metals:

1. In a metal, the valence electrons belong to the entire structure as itinerant electrons.
2. Interstices of the close-packed metal structure are taken as the location of electrons. Electron density is centered on an interstice. The binding regions for metal electrons may be considered to be the tetrahedral and octahedral interstices.
3. A preferential occupancy of octahedral or tetrahedral interstices depends on which interstice provides most effective screening from adjacent electrons by positive ion cores.
4. Electrons in adjacent interstices with minimum screening have opposite spins.
5. Before the number of electrons reaches 6, the d electrons are localized on the metal ion core.
6. Itinerant electrons in interstices act as ligands and determine the degeneracy of the d electrons localized on the ion core and, thus, their magnetic properties.
7. For metallic properties, vacant interstices must exist in the metal structure.
8. The shape of the octahedral or tetrahedral binding region can change with occupancy, since repulsion exerted by electrons in adjacent interstices defines the boundary of the binding region. Mutual polarization of itinerant electrons and ion cores may also modify electron density and "shape of the binding region."

These postulates imply that in an array of metal ion cores, the maximum attraction of these cores by electrons is obtained when some degree of localization of the electron density in the interstices of the structure exists. A consideration of the geometry of the various close-packed metal structures, as well as of the patterns of interstices within the structures, reveals that there are very important differences in interstice distances and arrangements. The theory gives an instantaneous picture in a dynamic situation of the electron density in tetrahedral and octahedral interstices, including both occupied and vacant interstices. The average itinerant electron density is accurately represented by the instantaneous picture, and a time average picture results when the total number of electrons in an interstice type is fractionally distributed over the total number of interstices of that type.

Other models, such as the orbital constructions of Trost [16] and the crystal-field band structures proposed by Goodenough [17], which we shall consider in Chapter X, emphasize the spatial arrangement of d electrons in transition metals. But the present model considers the itinerant electrons as ligands, not associated with a single metal ion core, and thus is related only indirectly to the models of Trost and Goodenough. Models of localized binding such as that of Pauling (Section 2.2) are much further removed from the band theory than the interstitial model. Application of the interstitial-electron model to chemisorption is speculative and qualitative at present, and we shall not pursue the model further. We note that the interstitial disposition of electrons could have significant implications for chemisorption, and the model is worth pursuing.

8.2 Electronic States in Chemisorption on Metals

No conclusions can be reached yet about the nature of a chemisorptive bond between a particular metal and foreign atom as to whether it is localized or nonlocalized. We note that the energy of the intermetallic bond in a crystal lies in the range 5–10 eV, the dissociation energy of a simple diatomic molecule is of the same order of magnitude, and the energy of the chemisorptive bond ranges from 2 to 10 eV. It is difficult to decide, therefore, whether to treat the surface interaction as a perturbation of the crystal by the approaching foreign atom, or to consider it a "surface compound" which is perturbed by the rest of the crystal. It seems probable a priori that examples of both types of systems should exist. First we shall discuss examples of those systems of chemisorption on metals which might reasonably be described as surface compounds. Similar, though less fundamental examples, were discussed in Section 2.3. After that, a discussion of treatments involving the entire metal plus foreign atom follows.

The Concept of a Surface Compound

The concept that a definite surface compound is formed in chemisorption is used in the interpretation of infrared spectra of molecules adsorbed on metals; for example, in the analysis of the linear and bridged forms of CO on nickel,

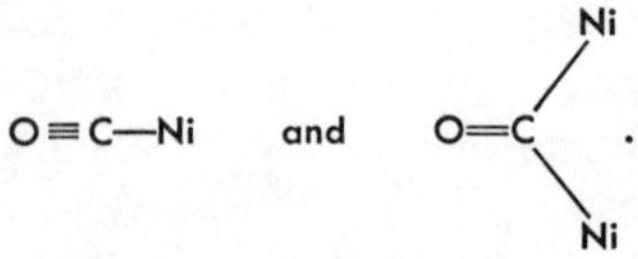

It is a consequence of the idea of "dangling" surface orbitals, which leads naturally to the formation of a surface compound. Yet, we must reconcile this model with the band theory of metals, which in many aspects has proven extremely useful. The strong overlapping of bands in metals places severe restrictions on the existence of discrete levels of surface compounds lying outside all of the metal band systems. Although the precise concept of a surface compound on a metal surface may run afoul of the band theory, it proves useful as an approximation. If d orbitals on a surface metal atom are coupled to orbitals of a foreign atom more strongly than to d orbitals on neighboring metal atoms, it is reasonable to adopt the concept of a surface compound perturbed by the rest of the metal. First, the discrete orbital energies of the surface compound would be calculated, and then allowances would be made for the interaction of the compound with the itinerant electron states of the surrounding metal. In the event that the interaction is weak, the energy levels of the surface compound that fall in the energy bands of the metal are merely shifted and broadened slightly. Originally discrete, they now become virtual levels lying within the energy bands of the metal. In this sense, the concept of a surface compound may be retained. As long as the interaction is weak, the energy of the virtual levels may be calculated approximately by assuming a surface compound. If there is no interaction at all between the surface compound and the rest of the metal, then the levels are discrete. But as we mentioned earlier, the probability of such levels decreases markedly with increasing overlapping of the energy bands of the metal. That a surface metal atom interacting with a foreign atom will become completely "demetalized" and lose its collective properties as it forms a nonmetallic molecule seems theoretically improbable, though definitive experiments have not been designed yet to give an unequivocal answer. If interaction between the surface compound and the itinerant electron states of the rest of the metal is strong, the levels are so broadened that they bear no resemblance to discrete levels, and the concept of a surface compound is worthless.

We see that our concept of a surface compound differs from the concept of an ordinary molecule in two ways. First, the metal atom of the surface compound is also a part of the bulk metal, and so itinerant electrons of

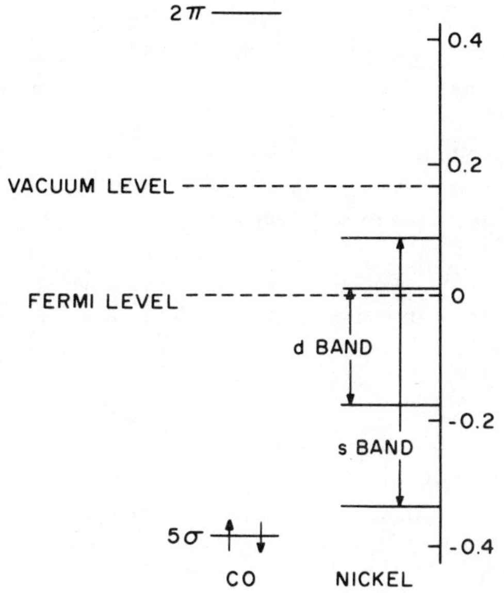

Fig. 8.2. The energy bands of nickel, and the 5σ and 2π levels o CO. Energy scale in hartrees (1 hartree = 27.21 eV).

the metal as well as electrons of the surface compound contribute to the bonding. Second, as a consequence of the coupling to the itinerant electron states, the surface compound may contain a nonintegral number of electrons. Only in the limit, when the wave functions of the surface compound drop to zero on all of the metal atoms except the central atom (or atoms) initially defined as part of the compound, does the surface compound become the exact counterpart of the simple, isolated molecule.

To give the concept of a surface compound a fair trial, a system with low heat of chemisorption should be selected. Grimley [4] has selected the system carbon monoxide chemisorbed on nickel, which turns out to be a good one. Other systems of low heat of adsorption are CO on iron, cobalt, rhodium, palladium, and platinum. Although an accurate theory of any of these systems is still remote, sufficient information exists for the nickel–carbon monoxide system to permit significant calculations. Accurate wave functions (Hartree–Fock type) and energies are available for carbon monoxide [18], and these may be combined with recent band structure calculations for nickel [19, 20]. We start by comparing the orbital energies of electrons in CO and nickel (Fig.8.2). The calculations of Matheiss give the positions of the s and d bands of nickel relative to the Fermi level, and the position of the Fermi level relative to the vacuum level is fixed by the work function. For our purposes, the position of the Fermi level may be considered independent of the crystal face, and is taken as the zero of energy. The energy of the first vacant level in CO [21], the 2π level, is $+0.44$ hartree (1 hartree = 27.21 eV). The highest filled level, 5σ, is separated from the 2π level by the very large energy gap 0.82 hartree. Because of this large separation, an electron placed in the 2π level would move in the self-consistent field of the other electrons without influencing them, and consequently an electron in this level does not give the negative ion CO^-. The doubly occupied 5σ level lies far below the Fermi level of nickel and the vacant 2π level lies far above it. For this reason, a bond involving significant sharing of electrons between the molecule and the metal should not form, and the binding energy should be quite small. According to Brennan and Hages [22], the binding energy of the nickel–CO bond is 0.067 hartree (1.82 eV or 42 kcal $mole^{-1}$), which is indeed weak. The system should provide a good test of the theory.

We summarize qualitatively the calculations which Grimley [4] has made on the nickel–CO system, using a simple linear model, Ni—C≡O. The 5σ orbital of CO and the $d\sigma$ orbital of the nickel atom are assumed to interact, forming the bonding and antibonding orbitals σ and σ^* of the compound. Similarly, the 2π orbitals of CO and the $d\pi$ orbitals of the nickel atom form the bonding and antibonding π orbitals. We describe the unperturbed σ and π orbitals of CO and the nickel atom as follows [23]:

1. The 5σ orbital of CO contains the unshared pair of electrons on the carbon atom.
2. The 2π orbitals of CO are the doubly degenerate antibonding π orbitals (empty).
3. The $d\sigma$ orbital of the nickel atom is the $d_z{}^2$ orbital.
4. The $d\pi$ orbitals of the nickel atom are the degenerate d_{xz} and d_{yz} orbitals.

In nickel metal, the electronic configuration of a nickel atom is $3d^{9.4}s^{0.6}$. As a good approximation, Grimley assumes that the $d\sigma$ orbital and the two $d\pi$ orbitals have energies at the Fermi level.

Before compound formation, each of the five d orbitals of a nickel atom in nickel metal have on the average 1.88 electrons, and thus the three orbitals involved in bonding have 5.64 electrons. The CO molecule has two electrons in the 5σ orbital and none in the 2π orbital, for a total of two electrons slated for bonding. After bonding, the numbers will change. Because of electron interactions, the energies of these orbitals depend on the energies relative to the Fermi level of the metal. Each orbital contains two electrons when it lies below the Fermi level, zero when above it, and

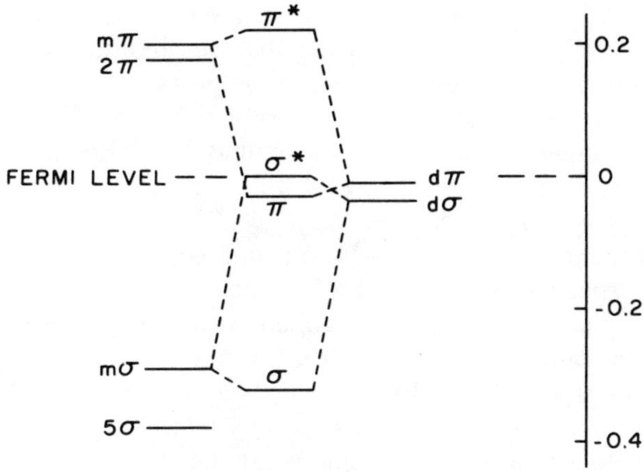

Fig. 8.3. The orbital energies of the surface compound formed when CO is adsorbed on nickel. The unperturbed dσ and dπ orbital energ'es coincide with the Fermi evel. Energy scale in hartrees (*4*).

between zero and two when it lies exactly at the Fermi level. Consequently, we see that the energies and the occupancies of the orbitals of the surface compound must be determined together self-consistently. Electron interaction in the surface compound is the result of electron interaction in the CO molecule and the metal atom separately, as well as the result of CO–Ni coupling. These changes upon compound formation, according to Grimley, cause a small shift of electrons from the nickel atom to the CO molecule, so that the CO molecule now has 2.284 electrons, an increase of 0.284, and the nickel atom has 5.426 in dσ and dπ orbitals, a decrease of 0.214. The final result of the quantum-mechanical calculations shows that the 5σ and 2π orbitals of an isolated CO molecule are raised when the surface compound is formed, and the dσ and dπ orbitals of the nickel atom are lowered. In the diagram in Fig. 8.3, these shifted levels are shown as mσ and mπ for CO and dσ and dπ for Ni, the latter having shifted downward from the Fermi level. The diagram indicates how these shifted levels interact to form the levels (σ and σ*) and (π and π*) of the surface compound.

So far, the calculations do not include the perturbation of the surface compound by the itinerant electron states of the metal. As a first approximation, Grimley assumes that only the sp band of nickel is involved in the coupling of the surface compound, for this band is much broader than the d band, whose energies lie within it. Once this coupling is triggered, all of the calculated discrete levels of the surface compound become virtual levels, and the number of electrons associated with the unperturbed surface compound are changed slightly. From quantum-mechanical theory,

Grimley concludes that such changes can only raise the total energy (<0) of the compound, which means that the binding of CO to nickel can only decrease as a result of including perturbations of the surface compound by the itinerant electron states of the metal. Since the binding energy of the unperturbed surface compound Ni–CO is calculated to be 0.0981 hartree, the binding energy of the perturbed compound will be slightly lower than that value.

Another localized approach has been proposed by Jansen [3]—a three-particle perturbation calculation for chemisorption on d-electron metals. Binding is assumed to occur through latent d-electron free valency of the metal ions, and the role of the conduction (itinerant electrons) of the metal is nonspecific and, like Grimley's approach, is treated as a perturbation. Jansen gives several reasons why chemisorption of a molecule on a single metal atom is not reasonable, among them the fact that chemisorption varies with crystallographic plane—a phenomenon which makes little sense if chemisorption is restricted to a single metal atom. The next simplest approximation is the consideration of the three-particle system, two metal ions and a chemisorbed particle. Moreover, Jansen selects the simplest three-particle system: two metal ions, each with one unpaired electron, and an adsorbate atom or molecule with two paired electrons. The theory predicts strong chemisorption of hydrogen molecules on a platinum surface, but cannot explain their dissociation on the surface. If a four-particle system is considered, the theory does predict dissociation [24]. The theory is not as well developed as Grimley's theory, and its value is difficult to assess at its present stage.

Accounts of other theories of localized adsorption may be found in Sections 2.3 and 3.3 related to covalent and ionic bonding, respectively, and in Section 7.6 concerning the use of small finite models. To these we should add a more detailed and specific calculation for hydrogen adsorption on a nickel crystal by Fassaert *et al.* [25]. These authors use the extended Hückel MO method and treat adsorption on Ni (111), (100), and (110). With a cluster of 13 nickel atoms having metallic properties, they found that the adsorption energy decreases with increasing number of neighbors, which bears a similarity to Grimley's [4] finding of highest adsorption energy when perturbations of the surface compound by itinerant electrons of the metal lattice are ignored.

Treatments Involving the Entire Crystal plus Foreign Atom

Ultimately, methods in this category should be the most rewarding, for in them is the latent power to decide for a specific system the nature of the chemisorption bond from the one extreme of discrete bonding to the other of metallic bonding. Yet enthusiasm must be tempered with an appreciation of the complexities. The simpler localized approaches of the preceding subsection will remain useful for some time to come.

We start with a discussion of calculations by Grimley, who treated the adsorption of an atom on nickel by the familiar LCAO self-consistent band picture, and extracted the concept of a surface molecule as the limiting case of the general theory. The treatment differs from his earlier work [4], described above, on the Ni–CO system, which starts with the assumption of a surface compound and modifies it by a small perturbation. Which of the two approaches to the concept of a surface molecule is to be preferred will depend on the metal–adsorbate system under consideration.

One of the specific systems considered by Grimley [26] is the chemisorption of an atom on the (100) face of nickel. He uses the value of 0.349 Ry (1 rydberg = 13.54 eV) from the work of Gerlach and Rhodin [27] for the work function of the clean surface, and this fixes the Fermi level, which is set at energy zero. Calculations on the free Na atom set the energy of the 3s valence orbital at -0.015 Ry below the Fermi level. At first thought, we would not expect the Na atom to lose its valence electron to the metal. But as we have already seen, the situation changes as the atom approaches the metal. The discrete atomic level is transformed into a virtual level with some broadening and with part of its density above the Fermi level. Thus the 3s orbital has a different degree of occupancy in the adsorbed atom (no occupancy above the Fermi level). A self-consistent calculation for the degree of occupancy and the orbital energy of the adsorbed atom must be made, and in this case Grimley uses the general approach, considering all the metal atoms plus the sodium atom. For this purpose, he employs a model Hamiltonian of the type first used by Anderson [28] in the theory of dilute alloys. It comprises four terms, two of which describe the free atom, one the metal, and the fourth the coupling between them. There are obvious objections to this Hamiltonian, such as its lack of terms for overlap between sodium and metal orbitals and its omission of certain terms for Coulomb interaction between electrons. But for qualitative initial studies, neglecting these terms is not serious. It is assumed that the chemisorbed sodium atom is positioned on four Ni surface atoms, giving a pyramidal Ni_4Na structure. Ultimately, in the general equations representing the entire system, only those terms involving interactions between the Na atom and its four nearest-neighbor Ni surface atoms will be retained. In this way, the concept of a surface compound is introduced. As in the case of the nickel–carbon monoxide system, which was attacked at the outset from the localized point of view, the surface compound is found to differ from an ordinary molecule, for it is an open system and need not contain an integral number of electrons.

Orbital energy and occupancy of the orbital of adsorbed Na were calculated as a function of the coupling parameter. In order to estimate a reasonable value of the coupling parameter, the experimental value of the heat of adsorption (0.2 Ry) was used. On this basis, the orbital energy of electrons in the surface bond was calculated to be −0.237 Ry.

Gadzuk *et al.* [29] have performed similar calculations for the chemisorption of alkali metal atoms on metals, using the Anderson model Hamiltonian. In considering the nature of the chemisorptive bond, they find the major portion of the binding is ionic. Values of binding energies in the range 1.5–2.5 eV were calculated, which, they cite, agree with experimental values for heats of chemisorption on tungsten, molybdenum, tantalum, and nickel. We note that Grimley [26] quoted an experimental value of 0.2 Ry (= 2.7 eV), which is in reasonable agreement with the range of experimental and theoretical values given by Gadzuk. A value of the orbital energy of electrons in the surface bond is not available for comparison with Grimley's value of 0.237 Ry (= −3.25 eV).

Grimley [30] has also studied the chemisorption of the hydrogen atom on metals. He considers first anionic adsorption, in which an electron is transferred from the metal to the atom, giving a negative hydrogen ion at the surface. The interaction energy Q (= $-\Delta H$) is given by an equation of the form

$$Q = -\phi + A_0 + K + J \tag{4}$$

where ϕ is the work function of the metal, A_0 the electron affinity of the hydrogen atom, and K and J the Coulomb and the quantum-mechanical exchange (resonance) interaction, respectively, of H^- with the metal. A simplified form of Eq. (4) neglects J and replaces K by the familiar mirror-image potential (see Section 3.2). The sum $(-\phi + A_0)$ is negative for all metals, which means that anionic interaction is endothermic, unless counteracted by $(K + J)$, and therefore not likely to occur. Grimley [30] sets up wave functions and derives energy expressions for K and J, representing interaction of H^- with the metal. He concludes that both terms involve only short-range interactions. Further, he presents evidence that the short-range interaction energies of such systems are negative. Thus the complete expression for Q in Eq. (4) is always negative and anionic chemisorption of hydrogen atoms is not favored. A similar conclusion was reached in Section 3.2, using the simplified form of Eq. (4) with the mirror-image potential in place of $(K + J)$.

Next, Grimley considers the formation of a covalent bond. He uses an

LCAO procedure similar to the one described in Sections 7.1 and 7.2, and comes to the same conclusion, namely, that provided surface states exist for the isolated metal, localized states exist for the combined system. If there is a localized state with energy lying below the Fermi level in the metal, the corresponding orbital will be occupied by two electrons of opposite spin, and something like a covalent bond will be formed between hydrogen and metal. As mentioned earlier, such a bond will be an open system, and in all probability will have a nonintegral number of electrons. Since a localized level must lie outside of all metal band systems, it is a difficult problem to determine just where, below the Fermi level, this localized state might be situated. A localized, covalent bond lying below the metal band structure should represent the important binding of a hydrogen atom to a metal surface. However, the existence of virtual levels at a higher energy level and within the band system, as described in the Ni–CO system, cannot be definitely excluded, though the binding energy would be less.

In a review article by Horiuti and Toya [31], two types of chemisorption of hydrogen atoms on metals, r and s, are distinguished, and LCAO-MO calculations made for each type. The r adatom is a chemisorbed hydrogen atom of conventional character, located outside the electron cloud of the metal surface and directly above a surface metal atom. The s atom is interstitially located between the electronic surface and the layer of surface metal atoms, nearly embedded in the electron cloud. They are illustrated in Fig. 8.4. The r adatom in its equilibrium position is slightly negatively polarized, and thus this type of adsorption increases the work function of the metal. The authors claim that electrical resistance measurements on nickel films with adsorbed hydrogen are in conflict with the concept of the formation of a localized bond, and they propose that the bond is metallic in nature.

The s adatom may be thought of as an atom dissolved in the metal and dissociated into a proton and an electron in the conduction band. The equilibrium position and the corresponding energy of the proton have been determined, using equations developed by Thomas [32], Fermi [33], and Weizsacker [34]. The energy in the equilibrium position was found to be lower than that of the dissolved state in the interior of the metal by 0.5 eV for nickel and 1.0 eV for platinum. This adatom is associated with a small positive dipole moment and is not bound to a particular metal atom. It may be viewed as the special case of a proton dissolved in the surface, and differs from adsorption models considered previously.

In concluding this section, we merely cite a paper by May and Carrol [35] that deals with ionic monolayers on metals and discusses the recon-

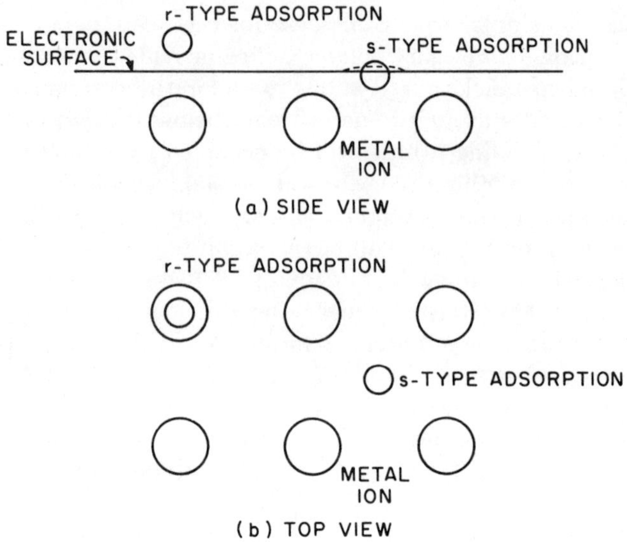

Fig. 8.4. The equilibrium of an r-type adatom above a metal ion, ~1 Å outside the electronic surface of the metal, and that of an s-type adatom interstitial, ~0.5 Å inside the electronic surface. (a) Side view; (b) top view (*31*). [After Horiuti and Tomiyuki, "Solid State Surface Science" (Mino Green, ed.), Vol. 1, Marcel Dekker, New York 1969. By permission of Marcel Dekker Publishing Company.]

struction of surfaces, and one by Grimley [36] that describes the chemisorption of a hydrogen atom on tungsten (100) as the surface complex W_2H.

8.3 Electron Densities

It was demonstrated some time ago [37, 38] that an impurity atom in a metal causes localized perturbations in electron density, long-range and oscillatory in nature. Grimley [39] speculated that a chemisorbed atom should produce similar perturbations, since it is an impurity atom on the surface. There are no experimental investigations of this phenomenon reported, and indeed, it is not yet clear exactly how to design a significant experiment. Grimley's speculations and calculations are important, for they point to interesting possibilities, which require corroboration or denial, and offer a challenge to experimentalists.

A simple LCAO–MO model is used, similar to the general model of Chapter VII, except that Wannier functions are employed instead of the more usual atomic wave functions. The wave function of an electron in a crystal may be expressed as a linear combination of Wannier functions which are approximately localized in the regions of the individual elementary cells (single atom per cell). A formal similarity exists between Wannier functions and the atomic orbitals of the LCAO–MO tight binding approximation. The electron density at a point is designated as $\rho(r)$ and is obtained by summing the square of the complete wave function, $|\psi(r)|^2$, for the crystal plus chemisorbed atom. If r is near a lattice point, we expect the main contribution to $\rho(r)$ from the wave function of the atom at that point. This approximation justifies the use of Wannier functions. Without the chemisorbed atom, the occupancy of metal orbitals should be the same at all lattice points in the metal surface. The presence of the chemisorbed atom disturbs this situation. Grimley has made calculations with and without chemisorption for several cases:

1. The system, crystal plus chemisorbed atom, has a virtual level near the Fermi level. (See Section 8.2 for the description of a virtual level associated with chemisorption.)
2. The system has a virtual level near the bottom of the conduction band.
3. The system has a discrete level.

In general, the calculations show that the disturbance is anisotropic. When R, the distance from a surface crystal atom, is in the critical direction, the disturbance falls away more rapidly than in other directions. For a body-centered cubic metal with a (110) surface, the critical direction is the (110) direction. Each of the three foregoing cases has a different expression for the electron density as a function of R, both for the critical direction and other directions, though approximately all vary as R^{-5} and R^{-3}, respectively. For the chemisorption of alkali metal atoms on tungsten or nickel, a virtual level near the Fermi level probably exists, and the critical direction is (110). On this basis, Grimley has calculated the change in occupancy, $\Delta n(R)$, of an orbital level on a surface metal atom at a distance R from the chemisorbed atom. The function $\Delta n(R)$ bears an obvious relationship to $\rho(r)$. In noncritical directions, the change in occupancy is given by

$$\Delta n(R) = 0.036(\sin 2.8R/R^3), \qquad R \text{ in Å}; \tag{5}$$

and in the critical direction,

$$\Delta n(R) = -0.059(\sin 2.8R/R^5). \tag{6}$$

Thus the simple model predicts a disturbance near a chemisorbed atom that is anisotropic, long range, and oscillatory.

The model has certain weaknesses, as do all simplifications of the complex real picture. In particular, Coulomb interactions need further study. Only Coulomb interactions between electrons in the chemisorbed atom were included, while those terms with both electrons in the metal, or one in the metal and one on the atom, were neglected. Problems associated with Coulomb interactions and effective shielding remain intractable in the theory of chemisorption. Until these problems are solved, no theory of chemisorption may be considered entirely satisfactory.

REFERENCES

1. I. Higuchi, T. Ree, and H. Eyring, *J. Amer. Chem. Soc.* **79,** 1330 (1957).
2. J. C. Robertson and C. W. Wilmsen, *J. Vac. Sci. Technol.* **8,** 53 (1971).
3. L. Jansen, *Mol. Proc. Solid Surfaces, Battelle Inst. Mater. Sci. Colloq., 3rd* p. 49 (1968).
4. T. B. Grimley, *Mol. Proc. Solid Surfaces, Battelle Inst. Mater. Sci. Colloq., 3rd* p. 299 (1968).
5. A. Clark, "The Theory of Adsorption and Catalysis," p. 141. Academic Press, New York, 1970.
6. M. Born and J. E. Mayer, *Z. Phys.* **75,** 1 (1932).
7. J. C. London, *Z. Phys.* **63,** 245 (1930).
8. J. C. Slater and J. G. Kirkwood, *Phys. Rev.* **37,** 682 (1931).
9. A. Mueller, *Proc. Roy. Soc. London* **A154,** 624 (1936).
10. D. R. Olander, *J. Phys. Chem. Solids* **32,** 2499 (1971).
11. O. Johnson, *Bull. Chem. Soc. Japan* **45,** 1599, 1607 (1972); *J. Res. Inst. Catal. Hokkaido* **19,** 152 (1972).
12. H. A. Lorentz, "Theory of Electrons." Teubner, Leipzig, 1906.
13. K. Fajans, *Ceram. Age* **54,** 288 (1949).
14. H. Bent, *J. Chem. Educ.* **42,** 348 (1965).
15. T. Berlin, *J. Chem. Phys* **19,** 208 (1951).
16. W. A. Trost, *Can. J. Chem.* **37,** 466 (1959).
17. J. B. Goodenough, *Phys. Rev.* **120,** 67 (1966).
18. W. M. Huo, *J. Chem. Phys.* **43,** 624 (1965).
19. L. F. Matheiss, *Phys. Rev.* **A134,** 970 (1964).
20. S. Wakok and J. Yamashita, *J. Phys. Soc. Japan* **19,** 1342 (1964).
21. R. C. Sahni, C. D. la Budde, and B. C. Sawney, *Trans. Faraday Soc.* **62,** 1993 (1966).
22. D. Brennan and F. H. Hages, *Phil. Trans.* **A258,** 347 (1965).
23. L. E. Orgel, "An Introduction to Transition-Metal Chemistry, Ligand-Field Theory." Wiley, New York and Methuen, London (1960).
24. A. van der Avoird, Thesis, Tech. Univ. Eindhoven, December, 1968.
25. D. J. M. Fassaert, H. Verbeek, and A. van der Avoird, *Surface Sci.* **29,** 501 (1972).
26. T. B. Grimley, *J. Vac. Sci. Technol.* **8,** 31 (1971).
27. R. Gerlach and T. N. Rhodin, "The Structure and Chemistry of Solid Surfaces" (G. A. Somorjai, ed.), p. 55. Wiley, New York, 1969.
28. P. W. Anderson, *Phys. Rev.* **124,** 41 (1961).

29. J. W. Gadzuk, J. K. Hartmann, and T. N. Rhodin, *Phys. Rev. B* **4,** 241 (1971).
30. T. B. Grimley, *in* "Chemisorption" (W. E. Garner, ed.). Butterworth, London and Washington D.C., 1957.
31. J. Horiuti and Toya Tomiyuki, "Solid State Surface Science" (M. Green, ed.), Vol. 1. Dekker, New York, 1969.
32. L. H. Thomas, *Proc. Cambridge Phil. Soc.* **23,** 542 (1927).
33. E. Fermi, *Z. Phys.* **48,** 73 (1928).
34. C. F. von Weizsacker, *Z. Phys.* **96,** 431 (1935).
35. J. W. May and C. E. Carroll, *Surface Sci.* **29,** 60, 85 (1972).
36. T. B. Grimley, *Ber. Bunsenges. Phys. Chem.* **75,** 1003 (1971).
37. J. Friedel, *Advan. Phys.* **3,** 446 (1954).
38. Yu. A. Izyumov, *Advan. Phys.* **14,** 569 (1965).
39. T. B. Grimley, *Proc. Phys. Soc. London* **92,** 776 (1967).

IX

Quantum Theory of the Chemisorptive Bond for Ionic Lattices

The problem of the degree of localization of a chemisorptive bond persists with ionic lattices. In simplified models, such as the LCAO–MO model, the parameters that distinguish localized from nonlocalized chemisorption are certainly well delineated and indicative of what might occur in real systems. But the problem is to determine the parameters for real systems so that the actual state of a chemisorptive bond might be spotted among the possibilities ranging from one extreme to the other. Though the problem remains stubborn, it is nevertheless instructive to consider the possibilities offered by theory. In Section 9.1, we describe the characteristics of two types of localized chemisorption states, Tamm states and Shockley states, as they might occur on a typical intrinsic semiconductor of the sphalerite type. They are similar in nature to the Tamm and Shockley states of free surfaces discussed in Section 6.1. Section 9.2 is concerned with ionic chemisorption, localized and nonlocalized, on semiconductors, and offers a more detailed quantum-mechanical interpretation of the boundary-layer theory, which was discussed in Section 5.2. The Madelung potential method, introduced in Section 6.4 in relation to free-surface states of ionic crystals, finds use also in the study of chemisorption states. Since the method is classical in nature, it gives only discrete levels. But an analog exists with quantum-mechanical methods. The discrete upper and lower bulk and surface levels associated with the M^+ and X^- ions may be considered to define a band gap in which are found the corresponding levels (or bands) of localized chemisorption states. In Section 9.3, we describe the use of the Madelung potential method superimposed upon the band picture to determine the nature of electronic states in chemisorption on ionic lattices. There follows a brief consideration of heats of chemisorption and their measurement in the light of the Madelung potential theory. With certain assumptions, theory and experiment are in good agreement for ionic chemisorption of oxygen and nitrous oxide on the $(11\bar{2}0)$ surface of CdS.

9.1 Tamm and Shockley States

Following Koutecky and Tomasek [1, 2], we use the sphalerite structure as the prototype of an intrinsic semiconductor. At sufficiently low temperatures, the crystal has a full valence band and an empty conduction band, for only an insignificant number of electrons possesses the energy to bridge the gap separating these bands. We consider localized chemisorption under these conditions. The crystal has two interpenetrating, face-centered lattices, one of cations and the other of anions. These lattices interpenetrate in such a way that each lattice cation is located at the center of a tetrahedron whose vertices are lattice anions. The central atom of the tetrahedron is assumed to have four sp^3 hybrid orbitals directed toward the vertices. Localized free-surface states are considered on the (111) face. The plane of this face cuts one localized bond between two neighboring atoms, forming either a cationic or an anionic surface. The surface is assumed to have a geometric structure identical to that of the bulk crystal. Self-consistent calculations are made using a one-electron Hamiltonian and the LCAO–MO approximation.

We assume that the crystal before chemisorption has free-surface states lying outside of the normal crystal band of energy states. These localized free-surface states may be Shockley states or Tamm states (Section 6.1). Shockley states arise when surface formation occurs through breaking of bonds between neighboring atoms in the crystal. The mere existence of a surface suffices; no perturbation of electrons or ions occurs at the surface. Shockley states are often called "dangling" bonds or free valencies, and their number is equal to the number of interrupted bonds on the surface. Tamm free-surface states exist if the potential in a surface cell differs from the potential inside the crystal. The mere existence of a surface is not enough; there must be a perturbation of the electron density at the surface. In the LCAO–MO approximation, the change in the potential at the surface is handled by using a different Coulomb integral (α') from that of the bulk (α).

We recall from Section 7.2 that all possible chemisorption states will be localized when free-surface states of the crystal alone exist. These states are assumed to form in this case by interaction of an adsorbate orbital and a free-surface orbital. They may be Shockley or Tamm states. Tamm chemisorption states exist if there is a difference in potential between the surface and the bulk after chemisorptive bond formation. The change of potential can occur in two ways: (1) by distortion of the surface atoms, and (2) by a difference in bond strengths between adsorbate–surface bonds and bonds within the crystal. Tamm chemisorption states occur only when the chemisorptive bond is stronger than the bonds in the bulk. Localization along this bond disappears if its strength decreases to that of bonds between crystal atoms. On the other hand, Shockley chemisorption states appear only if the chemisorptive bond is not too strong, or more precisely, when it is weaker than the bonds in the bulk crystal. It is evident that the type of chemisorption state, Tamm or Shockley, is not uniquely determined by the type of free-surface state entering into the interaction.

If the surface of the crystal has no localized states lying outside the bulk crystal bands, then localized chemisorption states may or may not exist, depending on the values of the critical parameters (Section 7.2). In this

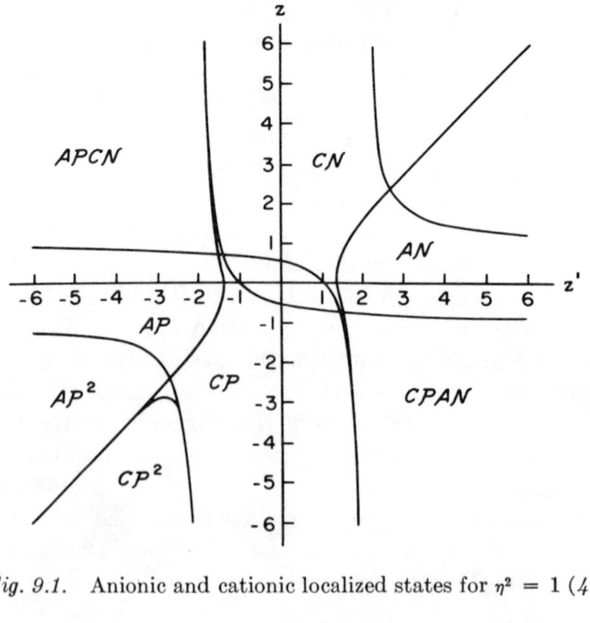

Fig. 9.1. Anionic and cationic localized states for $\eta^2 = 1$ (4).

case, the problem reduces to the Wolkenstein model [3], when localized states are postulated. According to this theory, a "weak" chemisorption may occur in which the chemisorbed particle and its adsorption center remain electrically neutral and in which the conduction electrons or holes of the lattice do not contribute to the chemisorptive bond. The bond is usually depicted as a one-electron bond. A "strong" chemisorption may also occur, in which either a conduction electron or hole participates in the bond. Such a bond contains two electrons and bears a charge. It can be formed only at temperatures sufficiently high that conduction electrons or holes are available. The nature of the free valencies, as Wolkenstein calls the conduction electrons or holes, is entirely different from that of the dangling bonds of Koutecky and Tomasek. In the following section, we give a rather detailed picture of localized and nonlocalized chemisorptive bonds formed by interactions of adsorbate particles and electrons from the conduction band or valence band in semiconductors.

9.2 Ionic Chemisorption†

In Section 5.1, we described the band theory of semiconductors and insulators, and in Section 5.2 we developed the boundary-layer theory of chemisorption on semiconductors, a classical theory superimposed on the band theory. This theory of chemisorption applies only to the purely ionic bond. We shall develop the quantum-mechanical picture of ionic chemisorption on semiconductors in more detail in this section, and shall not be limited to the purely ionic bond, but shall consider the entire range of bonds with ionic character. In addition to the usual types of cumulative and depletive chemisorption on n- and p-type semiconductors, listed in Table 5.1, other types, both localized and nonlocalized, will become evident. The procedure involves combining the results of Figs. 7.2 and 7.3 that give, as a function of the interaction parameters, the regions where localized chemisorption states occur and the curve for covalent states which separates anionic and cationic states. Both the nature of the adsorbed species and the way in which the balancing charge is carried in the crystal are determined.

Consider a perfect crystal with N electron levels, and bring up a hydrogenlike atom having a single s electron in its outermost orbital. The combined system has $N + 1$ levels, each capable of accommodating two electrons. Assume that before chemisorption each crystal orbital in the valence band is doubly occupied. The foreign atom brings up only one

† Ref. [4].

electron involved in chemisorption. Therefore, one of the $N + 1$ orbitals will be singly occupied. Since the electron levels are filled from the lowest level up, the highest level will contain only one electron. If the energy of the chemisorption state lies below the highest level, it will be doubly occupied, and the singly occupied level will not be of primary concern. We are also interested in the electronic charge distribution among the $N + 1$ atoms of the system. Each energy level has an associated wave function ψ_j which is expressed in the LCAO approximation as the sum of the atomic orbitals ϕ_r, one for each atom of the system including the foreign atom, multiplied by appropriate coefficients,

$$\psi_j = \sum_r C_{rj}\phi_r. \tag{1}$$

When the integrals $\int \psi_j^2 \, d\tau$ and $\int \phi_r^2 \, d\tau$ are normalized and the overlap integrals $\int \phi_r \phi_s \, d\tau$, $r \neq s$, set equal to zero, the square of the coefficient C_{rj} is interpreted as the probability that the charge of the electron with energy E_j will be located on the rth atom. If we sum the squares of the coefficients of the rth atom over all energy levels of the system, we obtain the total charge q_r on that atom,

$$q_r = 2 \sum_j C_{rj}^2, \tag{2}$$

where the factor of 2 is used because each level is doubly occupied. Quantum-mechanical calculations show that $q_r = 2$, that is, $\sum C_{rj}^2 = 1$ for all r. In other words, all atoms have the same total charge. For instance, if the chemisorptive bond were localized and strongly ionic, then one of the members of the bond would contain a dearth of charge which would be made up to a total charge of 2 by nonlocalized states of the system. In summary, we find two electrons per state and a total charge of 2 on each atom of the system. We can draw the conclusion that a hydrogenlike atom with a charge of 2 (excluding inner electrons) will exist on the surface as an anion. A few examples will clarify the picture.

Anionic Chemisorption on p-Type Semiconductors

The arguments presented above for a perfect crystal apply if the crystal is a p-type semiconductor either because it contains foreign impurities or is nonstoichiometric. But the number N must refer only to the levels in the normal crystal band and not to the discrete levels lying above it which are associated with p-type semiconductivity. Similar arguments apply to n-type semiconductors.

By anionic chemisorption, we mean that the distribution of charge between the chemisorbed atom and the crystal leaves an excess negative charge on the foreign atom. We want to know in what regions of Fig. 7.2 this occurs and how the balancing positive charge is carried in the crystal. Interaction is assumed to occur between the orbital on the foreign atom and the normally filled valence band of the crystal. Situated just above the valence band are the discrete levels that give p-type semiconductivity, and they are normally empty. In Fig. 9.1, we have superimposed Fig. 7.2, showing the occurrence of localized chemisorption states, on Fig. 7.3, showing the occurrence of covalent $\mathcal{P}$ states. The curve for occurrence of covalent $\mathcal{N}$ states has been added in Fig. 9.1. The covalent curves ($R = \frac{1}{2}$) separate the $\mathcal{P}$ and $\mathcal{P}^2$ regions into anionic (A$\mathcal{P}$, A$\mathcal{P}^2$) and cationic (C$\mathcal{P}$, C$\mathcal{P}^2$) areas, and similarly for the $\mathcal{N}$ and $\mathcal{N}^2$ regions, while the $\mathcal{PN}$ regions become A$\mathcal{P}$C$\mathcal{N}$ and C$\mathcal{P}$A$\mathcal{N}$ areas. The forbidden region, where no localized states exist, is unchanged. Results for the various types of anionic chemisorption on p-type semiconductors are summarized in Table 9.1. We recall from Section 7.1 that $\mathcal{P}$ and $\mathcal{N}$ states are localized chemisorption levels. With no loss in generality, we consider that $\mathcal{P}$ states exist below the bottom edge of the valence band and $\mathcal{N}$ states above the top edge, though the reverse is possible, depending on the sign of the resonance integral β (Section 7.1). In the second column of the table, "between" means that the $\mathcal{N}$ level lies between the top edge of the valence band and the discrete impurity level, and "above" means that the $\mathcal{N}$ level lies above the impurity level. The third column shows how the balancing positive charge is carried in the crystal. A positive trapped hole occurs in the $\mathcal{N}$ level, which is singly occupied because it is the highest level of the system. The fourth column of the table describes chemisorption as cumulative if it is not limited by the number of impurity levels in the solid and as depletive if it is. Other types of anionic chemisorption are possible on p-type semiconductors, but these are the important ones.

TABLE 9.1

Anionic Chemisorption on p-*Type Semiconductors*[a]

Region	Level	Balancing charge	Chemisorptive type
$\mathcal{P}$, $\mathcal{P}^2$	—	Free holes	Cumulative, Type A
C$\mathcal{N}$, A$\mathcal{P}$C$\mathcal{N}$	Between	Trapped hole	Cumulative, Type B
Forbidden	—	Free hole	Cumulative, Type C
C$\mathcal{N}$, A$\mathcal{P}$C$\mathcal{N}$	Above	Trapped hole and occupied impurity levels	Depletive

[a] Grimley [4].

TABLE 9.2

Wave Function Coefficients for an Anionic $\mathcal{P}$ State[a]

r	λ	0	1	2	3	4	5
C_r	3.160	1.000	0.316	0.100	0.032	0.010	0.003

[a] From Grimley [4]; $z = 0$, $z' = -3.16$, $\eta = 1$; coefficient unnormalized.

The usual picture of anionic chemisorption on p-type semiconductors, described in the boundary-layer theory (Section 5.2), is closely related to what happens in the areas designated A$\mathcal{P}$ and A$\mathcal{P}^2$ in Fig. 9.1. The foreign atom brings one extra level to the system, but only one electron. Thus the combined system has one level singly occupied, and this will be the highest level. Since the $\mathcal{P}$ level is below the bottom edge of the valence band, it will be doubly occupied. The extra electron comes from a nonlocalized level, leaving a free hole in the crystal, one for each foreign atom chemisorbed. The charge distribution produces an anion on the surface, as described above, and in the A$\mathcal{P}$ and the A$\mathcal{P}^2$ areas the greater part of this charge is supplied by the localized electrons in the $\mathcal{P}$ state. Grimley calculates the first few wave function coefficients for the anionic $\mathcal{P}$ state with $z = 0$, $z' = -3.16$, and $\eta = 1$, and the results are shown in Table 9.2. The value of R, the charge order (Section 7.4) for the foreign atom produced by a single electron in the $\mathcal{P}$ state, is $R = 0.900$, which means that the electron is well localized on the foreign atom. Since the $\mathcal{P}$ state is doubly occupied, it contributes 1.8 to the total charge of 2 on this atom. Consequently, the nonlocalized states contribute only 0.2.

Cumulative chemisorption, Type A, is also found in areas C$\mathcal{P}$ and C$\mathcal{P}^2$. Although the localized $\mathcal{P}$-state bond is cationic, the total charge on the foreign atom remains 2, so that the foreign atom exists as an anion. A calculation similar to that for an anionic $\mathcal{P}$ state is shown in Table 9.3. The charge order on the foreign atom from the localized $\mathcal{P}$ state is 0.066, which shows that the wave function is well localized on the first crystal atom. With the $\mathcal{P}$ state doubly occupied, the localized electrons contribute only 0.132 to the total charge of 2 on the foreign atom. Hence, the remainder of the charge, 1.868, is contributed by nonlocalized electrons of the system located at higher energy states. This picture differs from the usual one depicted by the boundary-layer theory, where the foreign atom accepts an electron from the solid into its valence orbital. We see that cumulative chemisorption as anions on p-type semiconductors may occur throughout the $\mathcal{P}$ and $\mathcal{P}^2$ regions. The examples given are rather extreme cases. On the covalent line between the A𝒫 and C𝒫 areas, a localized homopolar bond exists between the foreign atom and the crystal. Since $R = \frac{1}{2}$ on this line, the charge on the foreign atom is contributed equally by the two electrons forming this bond and those in the localized states of the system. A second type of cumulative chemisorption (C) may occur when no localized states exist. In this case, interaction occurs in the forbidden zone of Fig. 9.1.

A third type of anionic chemisorption is cumulative Type B, listed in Type 9.1. It involves a cationic 𝔑 level which lies between the top edge of the valence band and the discrete impurity levels. Since it is the highest normally occupied level in the system, it contains only one electron, and the vacancy in this level is the trapped positive hole. The hole is trapped in the neighborhood of the first crystal atom because the wave function of a cationic level is localized on this atom. If the hole is well localized, the electron missing from the C𝔑 level does not greatly affect the charge on the foreign atom, which therefore exists as an anion on the surface. In other words, taking a positive hole from the foreign atom, when its orbital is transformed upon interaction with the crystal into a C𝔑 orbital which is centered about the first crystal atom, is tantamount to transferring a negative charge to the foreign atom from the crystal atom. At low temperatures, the conductivity of the solid will not be affected, for the valence band remains full. As the temperature is increased, electrons are excited to the 𝔑 level from the valence band and free positive holes produced. The nonlocalized electrons give the anionic charge to the foreign atom. Although chemisorption of this type has been proposed for oxygen on cuprous oxide [5], its occurrence is probably rare because of the requirement that the cationic 𝔑 level lie in the narrow region between the top edge of the valence band and the discrete impurity levels.

A fourth possible type of anionic chemisorption on p-type semiconductors is interesting because it is depletive. In this case, the cationic 𝔑 level is presumed to lie above the discrete impurity levels. The electron which the foreign atom brings to the system goes into one of these levels.

TABLE 9.3

Wave Function Coefficients for a Cationic 𝒫 State[a]

r	λ	0	1	2	3	4	5
C_r	0.278	1.000	0.305	0.093	0.028	0.009	0.003

[a] From Grimley [4]; $z = -3$, $z' = 0$, $\eta = 1$; coefficients unnormalized.

As in the previous case, the $C\mathcal{M}$ level is localized on the first crystal atom, which in this case contains two trapped positive holes because the single electron brought in by the foreign atom has dropped into a discrete impurity level. For each foreign atom chemisorbed, two positive holes are trapped in the vicinity of the first crystal atom, resulting in conversion of the foreign atom into an anion. Chemisorption is limited by the concentration of impurity levels into which the single electron on each foreign atom falls, and is therefore depletive.

Anionic chemisorption is favored over the metal cations, with interaction taking place between the orbitals on the foreign atom and the cation band, for example, the 3d band in Cu_2O. If this is the highest filled band, the mechanisms of this section are most likely to occur.

Cationic Chemisorption on p-Type Semiconductors

According to the boundary-layer theory, cationic chemisorption on a p-type semiconductor occurs by transfer of an electron from the foreign atom to an impurity level in the solid lying just above the highest filled band. The foreign atom is converted to the cation, and the chemisorption is depletive, for only a limited number of impurity levels exist. Such chemisorption may occur in the $A\mathcal{M}$, $A\mathcal{M}^2$, and $C\mathcal{P}A\mathcal{M}$ areas, provided the $\mathcal{M}$ level lies above the impurity levels. The $A\mathcal{M}$ chemisorption state is vacant because the electron from the foreign atom has gone to the impurity level. The wave function of an $A\mathcal{M}$ state will be centered about the foreign atom, so that the positive hole will be trapped by that atom, making it a cation.

In oxide semiconductors, cationic chemisorption should occur on the crystal anions, while the electron lost by the foreign atom should interact on the impurity level of a crystal cation. Thus, the interaction involves both anion and cation bands. Such a process is extremely complex, and no model proposed for metal oxides—especially transition metal oxides, which are of greatest importance in catalysis—is sufficiently adequate.

Anionic Chemisorption on n-Type Semiconductors

Here the usual picture, according to the boundary-layer theory, is that the foreign atom accepts an electron from an impurity level of the crystal, thus becoming an anion. For an n-type semiconductor, impurity levels exist just below the conduction band and are normally full. Since the concentration of impurity levels is very low, chemisorption of this type is depletive. For the quantum-mechanical picture, it is assumed that interaction occurs between the orbital on the foreign atom and the conduction band of the crystal. Such interactions, closely analogous to those assumed in the boundary-layer theory, may occur in the $A\mathcal{P}$, $A\mathcal{P}^2$, and $A\mathcal{P}C\mathcal{N}$ areas of Fig. 9.1, provided the $\mathcal{P}$ level lies below the impurity levels. With semiconducting oxides, anion chemisorption should take place over lattice cations, and the simple molecular orbital picture is adequate if the conduction band is associated with the cation lattice. There is direct evidence that this is true for Al_2O_3, for soft X-ray emission spectra show that the highest filled band is the oxygen 2p band [6].

Other areas of Fig. 9.1 may be associated with anionic chemisorption on n-type semiconductors. Indeed, if interaction is assumed between the orbital on the foreign atom and the highest filled band of the crystal, anionic chemisorption is found in all areas of the figure, provided the highest localized level, $\mathcal{N}$ or $\mathcal{P}$, falls below the impurity levels in the solid.

Cationic Chemisorption on n-Type Semiconductors

The boundary-layer theory tells us that the foreign atom loses an electron to the conduction band of the crystal. Chemisorption is cumulative, conductivity of the solid is increased, and the electron which the foreign atom brings to the system is nonlocalized. Since interaction is with the conduction band and the foreign electron is nonlocalized in it, chemisorption of this type must occur in the forbidden region of Fig. 9.1. The amplitude of the wave function of the nonlocalized electron is very small on the foreign atom, which therefore exists as a cation on the surface.

Molecular orbital theory predicts other types of cationic chemisorption on n-type semiconductors. For example, cationic chemisorption may occur in the $C\mathcal{P}$, $C\mathcal{P}^2$, and $C\mathcal{P}A\mathcal{N}$ areas, provided the $\mathcal{P}$ level lies below the impurity levels. An electron is lost from the normally full impurity level for each atom chemisorbed, and two electrons are trapped in the $C\mathcal{P}$ level, one from the impurity level and one from the foreign atom. The wave functions for these electrons are large on the first crystal atom and small on the foreign atom, which exists, therefore, as a cation. In this case, chemisorption is depletive. However, if the $C\mathcal{P}$ level lies between the conduction band and the impurity levels, chemisorption is still cationic, but the electrons in the impurity levels do not enter into the interaction, and only one electron is trapped in the $C\mathcal{P}$ level in the vicinity of the first crystal atom. Chemisorption is cumulative.

Other variations are possible, as a little reflection will verify, but the few cases described should suffice to indicate the broad scope of chemisorption systems which theory predicts.

9.3 The Madelung Potential Method†

ELECTRONIC STATES

The use of the Madelung potential method for determining the bulk and free-surface electronic states of ionic crystals was described in Section 6.4. We shall extend it, following Mark [7, 8], to the electronic states of chemisorbed species on ionic crystals. The method is classical in nature, giving only discrete levels, but it may be interpreted in the light of quantum-mechanical theory. For convenience, the important equations for bulk and free-surface states will be summarized here and then used in deriving equations for chemisorption states.

The energies of the bulk cationic and anionic levels are given, respectively, by

$$E_{\mathrm{M}} = (CZe/r_0) - I_{\mathrm{M}}, \tag{3}$$

and

$$E_{\mathrm{X}} = -(CZe/r_0) - A_{\mathrm{X}} \tag{4}$$

where I_{M} and A_{X} are the ionization potential of the cation M and the electron affinity of anion X, respectively; Z is the ionic valence and may be fractional to account for covalent contributions; e is the charge on an electron; and r_0 is the lattice constant. The Madelung constant C, which is the same for cation and anion if the lattice contains an equal number of each, is expressed by

$$C = \sum_{i,j,k} q_{i,j,k}/R_{i,j,k} = \sum_{i,j,k} Q \tag{5}$$

where $q_{i,j,k} = \pm 1$, depending on the sign of the ion located at the space index position (i, j, k), and $R_{i,j,k}$ is the distance measured from $(0, 0, 0)$ to (i, j, k) expressed in units of the lattice constant. Each index runs from $-\infty$ to $+\infty$. The sum in Eq. (5) may be broken down into the sums for all points on a plane through the crystal at $i = 0$, for all points above; and for all points below this plane $(i > 0, i < 0)$,

$$C = \sum_{i=0} Q + \sum_{i>0} Q + \sum_{i<0} Q = \sum_{i=0} Q + 2 \sum_{i>0} Q, \tag{6}$$

where the second equality is the result of symmetry above and below the plane $i = 0$. The indices j and k are understood to extend over all space in the directions parallel to the plane $i = 0$. When the crystal is cut on the plane $i = 0$, a surface is produced. The Madelung constant of a surface

† Refs. (7, 8).

ion will extend over the region $i \geq 0$, and is written

$$C_s = \sum_{i=0} Q + \sum_{i>0} Q. \tag{7}$$

Combining Eqs. (6) and (7), we obtain

$$C_s = \tfrac{1}{2}C + \tfrac{1}{2} \sum_{i=0} Q. \tag{8}$$

Energy equations for surface states are now obtained identical to Eqs. (3) and (4) except that C, the bulk Madelung constant, is replaced by C_s, the surface Madelung constant. We note that the quantity

$$\Delta V = (CZe/r_0) \tag{9}$$

of Eqs. (3) and (4) is the interaction potential in the classical limit (overlap and resonance integrals ignored) and approaches zero at very large values of r_0.

We now consider an ion adsorbed on the surface designated by the plane $i = 0$. Near the ionic crystal surface, it experiences a shift in its potential, either electron affinity or ionization potential, caused by electrostatic interaction with the ionic lattice. The magnitude of this shift is estimated by calculating the Madelung constant C_a at the position of the adsorbed ion. To simplify the calculations, the adsorbate site is taken directly above a lattice site and at one lattice constant distant from it. The constant C_a is obtained from Eq. (7) by omitting the sum of the plane $i = 0$,

$$C_a = \sum_{i>0} Q. \tag{10}$$

This is equivalent to removing all ions of the surface plane except one, thus wiping out all interactions on that plane. From Eqs. (7), (8), and (10), we obtain

$$C_a = C - C_s. \tag{11}$$

Corresponding to ΔV in Eq. (9), we have

$$\pm\Delta V_a = Z_a C_a e/r_0 \tag{12}$$

where the positive sign is used if the adsorbed ion is located above a negative lattice site, and the negative sign if it is above a positive lattice site. As before, the ionic valence Z_a may be fractional to account for covalent contributions. The value of the interaction potential may have any value between $\pm\Delta V_a$, depending on the location of the adsorbed ion; for example, midway between two lattice sites, $\Delta V_a = 0$. For a single adsorbate

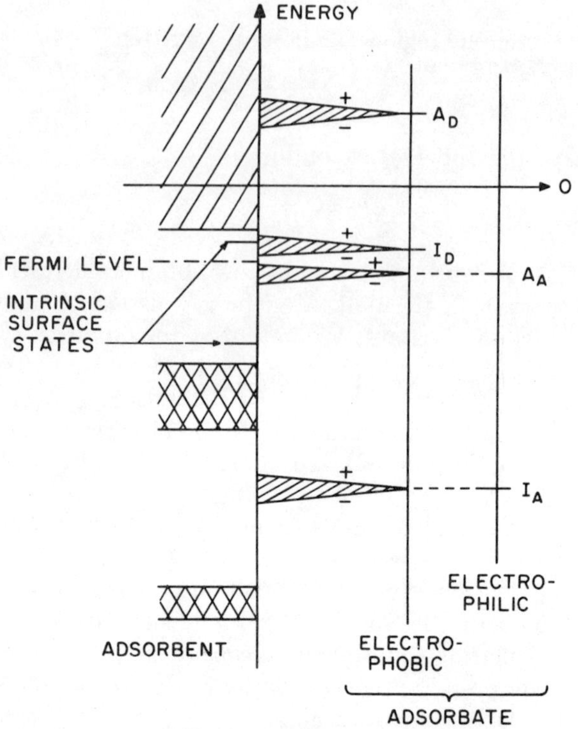

Fig. 9.2. Potential energy diagram illustrating the evolution of the chemisorption states on an ionic surface. On the left side is the energy band diagram of the adsorbent. Shown are two valence bands, the conduction band, the two intrinsic surface states of the principal gap, and the Fermi level. On the right side are the potential energy diagrams of a donorlike electrophobic adsorbate and of an acceptorlike electrophilic adsorbate. The range of adsorbate levels on the surface is indicated by the shaded regions; the branches labeled − and + refer to the adsorbate situated on cationic and anionic lattice sites, respectively. [After Mark [7]. By permission of the Pergamon Publishing Company.]

ion, ΔV_a may be interpreted as a correction to the ionization potential or electron affinity as the ion approaches the surface.

Chemisorption is illustrated in Fig. 9.2 as it progresses from the isolated states of crystal and adsorbate to the combined state. The left side of the diagram depicts the energy band diagram of the crystal; it includes two well-separated and filled valence bands, the normally empty conduction band, two free-surface states, and the Fermi level. On the right side are the potential energy levels of an electrophilic and an electrophobic adsorbate. An electrophobic adsorbate has a low ionization potential and a positive electron affinity; an electrophilic adsorbate has a high ionization potential and a negative electron affinity. The diagram also gives the range of adsorbate levels on the surface; the branches labeled − and + indicate the adsorbate located directly above a cation lattice site and an anion lattice site, respectively. As mentioned earlier, all intermediate states are possible. An adsorbate is donorlike when its corrected ionization potential, that is, the energy of its unoccupied level in the combined system,

$$E_D = -I_D \pm \Delta V_a, \tag{13}$$

lies near or above the Fermi level. An adsorbate is acceptorlike when its corrected electron affinity, that is, the energy of its occupied level in the combined system,

$$E_A = -A_A \pm \Delta V_a, \tag{14}$$

lies near or below the Fermi level. Electrophobic adsorbates (such as alkali metals) are more apt to be donorlike, while electrophilic adsorbates (such as oxygen and halogens) tend to act as acceptors.

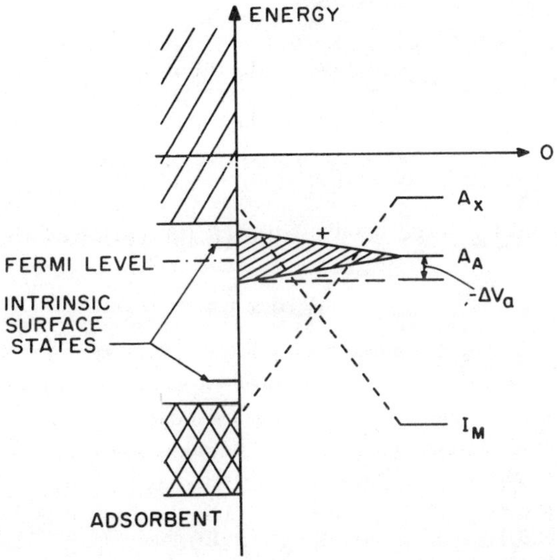

Fig. 9.3. Potential energy diagram of an electrophilic, acceptor-like adsorbate on an ionic adsorbent. I_M and A_X are, respectively, the free-space ionization potential of the cationic lattice component and the electron affinity of the anionic component. A_A is the free-space electron affinity of the adsorbate assumed more negative than A_X. [After Mark [7]. By permission of the Pergamon Publishing Company.]

Equations (13) and (14) give the energy of the chemisorption state of an adsorbed cation and an adsorbed anion. We wish to locate these energies with respect to the bottom of the conduction band. The specific case of an acceptorlike adsorbate is considered, whose electron affinity ($-A_A$) in the isolated state is more negative than that of the anionic lattice constituent ($-A_X$). The situation is shown in Fig. 9.3 and is representative of oxygen chemisorbed on CdS. In the classical limit, the energies of the conduction band (E_M) and the highest valence band (E_X) are given by Eqs. (3) and (4), respectively. The gap between the conduction band and the valence band is

$$E_M - E_X = \Delta E_g = 2\Delta V - (I_M - A_X). \tag{15}$$

The depth ΔE_a of the adsorbate level below the conduction band is

$$E_M - E_A = \Delta E_a = \Delta V \pm \Delta V_a - (I_M - A_A). \tag{16}$$

Using Eq. (15), the depth ΔE_a may be expressed by

$$\Delta E_a = \frac{\Delta E_g}{2}\left[\frac{1 \pm (Z_a/Z)(1 - \Gamma) - (2Z_a/Z)\mu_a}{(1 - \mu)}\right] \tag{17}$$

where

$$\mu = (I_M - A_X)r_0/2ZCe,$$

$$\mu_a = (I_M - A_A)r_0/2Z_aCe,$$

and

$$\Gamma = C_s/C.$$

Values of Γ and μ have been evaluated for various halides, oxides, and

TABLE 9.4

Γ *Values of Several Ionic Surfaces*[a]

Lattice type	Surface index	Γ
NaCl	(100)	0.96
NaCl	(110)	0.86
NaCl	(210)	0.77
NaCl	(211)	0.60
CsCl	(110)	0.90
Wurtzite	$(11\bar{2}0)$	0.88
Wurtzite	$(10\bar{1}0)$	0.79
Zinc blende	(110)	0.85

[a] Mark [7].

TABLE 9.5

Values of μ for Several Halides, Oxides, and Sulfides[a]

Material	Lattice type	($Z = 0.5$)	($Z = 1$)	($Z = 2$)
Alkali halides	NaCl		0.12	
AgF	NaCl		0.16	
AgCl	NaCl		0.21	
AgBr	NaCl		0.21	
AgI	Zinc blende		0.28	
MgO	Wurtzite	0.09	0.23	0.50
BaO	NaCl		0.16	0.52
NiO	NaCl	0.06	0.23	0.57
ZnO	Wurtzite	0.17	0.30	0.56
CdO	NaCl	0.15	0.31	0.60
ZnS	Zinc blende	0.33	0.40	0.57
CdS	Wurtzite	0.33	0.41	0.58

[a] Mark [7].

sulfides [9], and representative values for compounds with equal numbers of positive and negative ions are given in Tables 9.4 and 9.5. Three values for μ are listed in many cases (Table 9.5) because of the uncertainty about the ionic valency Z. In a broad review of experimental results, Birman [10] concludes that the best value of Z is near 0.5 for the oxides and sulfides with wurtzite and zinc blende structures, and somewhat larger for cubic lattices.

Mark [7] has considered the case of oxygen chemisorbed on the $(11\bar{2}0)$ face of CdS, using $Z = 0.5$ (62.5% ionicity), which fixes the value of μ. He also takes μ_a from the data for CdO in Table 9.5 at $Z = 0.5$, and further assumes that $Z_a = 0.5$. With these values ΔE_a ranges from 1.04 eV to 1.47 eV, the larger value referring to oxygen located on a Cd site. The value of Z_a should be taken larger than Z, since CdO is cubic and, therefore, the Cd–oxygen bond probably has a greater ionicity than the Cd–sulfur bond. Experiment [11] gives a value of $\Delta E_a = 0.9$ eV, which corresponds to a value of $Z_a \cong 0.75$ (69% ionicity) for oxygen chemisorbed on a Cd site and to a value of $Z_a \cong 0.55$ (63% ionicity) for oxygen chemisorbed on a sulfur site. Oxygen is more likely to adsorb nearer a Cd site, so that $Z_a \cong 0.75$ is the more probable value. Estimates have also been made for oxygen chemisorbed on ZnO and ZnS, again using $Z = 0.5$. Results of estimates for four Z_a values on stable surfaces are summarized in Table 9.6, including available experimental data. Agreement is satisfactory when Z_a is chosen close to the corresponding lattice valence, which is reasonable.

TABLE 9.6

Chemisorption State of Oxygen on CdS, ZnO, and ZnS[a]

A. CdS: $\Delta E_g = 2.4$ eV at room temperature,[b] (11$\bar{2}$0) surface

		ΔE_a (eV)		
		Chemisorbed on		
Z_a	Ionicity (%)	Cd site	S site	Exp. value
0.5	62.5	1.47	1.04	
0.75	68.75	0.89	0.24	0.91 ± 0.03[c]
1.0	75	0.00	−0.86	
2.0	100	−5.96	−7.67	

B. ZnO: $\Delta E_g = 3.2$ eV at room temperature,[b] (11$\bar{2}$0) surface

	ΔE_a (eV)		
	Chemisorbed on		
Z_a	Zn site	O site	Exp. value
0.5	1.50	1.04	
0.75	0.91	0.00	1.05 ± 0.05[d]
1.0	0.08	−0.85	
2.0	−5.80	−7.65	

C. ZnS: $\Delta E_g = 3.7$ eV at room temperature,[b] (110) surface

	ΔE_a (eV)		
	Chemisorbed on		
Z_a	Zn site	S site	Exp. value
0.5	2.23	1.41	
0.75	1.43	0.19	No data available
1.0	0.28	−1.38	
2.0	−7.95	−11.3	

[a] From Mark [7]; lattice valence $Z = 0.5$ [10].
[b] Bube [12].
[c] Mark [11].
[d] Watanabe *et al.* [13].

All of the results indicate that the adsorption bond should be considered as partially ionic with some covalent character. The possibility of a doubly charged chemisorbed oxygen is considered remote. Negative values of ΔE_a mean that the chemisorption state is located above the edge of the conduction band, which is compatible with an appreciable covalent contribution.

Heats of Chemisorption

We see in Table 9.6 that the chemisorption states of oxygen on CdS and on ZnO lie about 0.91 eV and 1.05 eV, respectively, below the conduction band. Equation (16) defines ΔE_a, and it is called the desorption energy. Unlike the more complex quantum-mechanical theories of chemisorption, the Madelung potential theory is expressed in terms that can be measured experimentally. We describe qualitatively a method for measuring ΔE_a [11]. For further details, the review article by Mark [8] may be consulted.

The method is based on the illumination of an ionic solid by light of greater energy than the band gap, thus exciting electrons across it. In Fig. 9.4a, the steady-state model of chemisorbed ions during illumination is shown [8] with the adsorbate level located between the upper and lower free-surface states. The depletion layer, containing trapped holes (+), compensates for the electrons trapped in either free-surface states or the adsorbate level. The time for holes to drift to the surface is assumed to

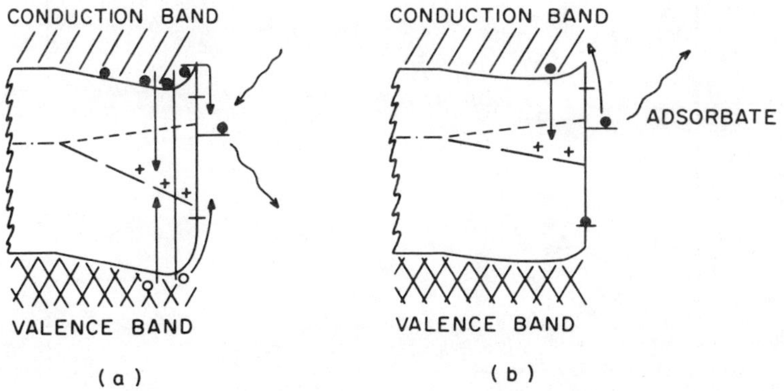

Fig. 9.4. (a) Steady-state band diagram for the formation of chemisorbed ions during illumination. (b) Band diagram during the photoconductivity decay in the presence of adsorbate that commenced at the state shown in part (a). [After Levine and Mark [8]. By permission of the American Institute of Physics.]

be less than the lifetime of a hole. It is evident that adsorbed ions trap conduction band electrons, thus causing a decrease in photocurrent at steady state. $E_{f,n}$ and $E_{f,p}$ are the steady-state quasi-Fermi levels for electrons and holes, respectively, and E_f is the equilibrium Fermi level of the undisturbed bulk crystal. When illumination is switched off, the free carriers in the bulk (positive holes and electrons) recombine and the quasi-Fermi levels recede from their respective band edges, ultimately

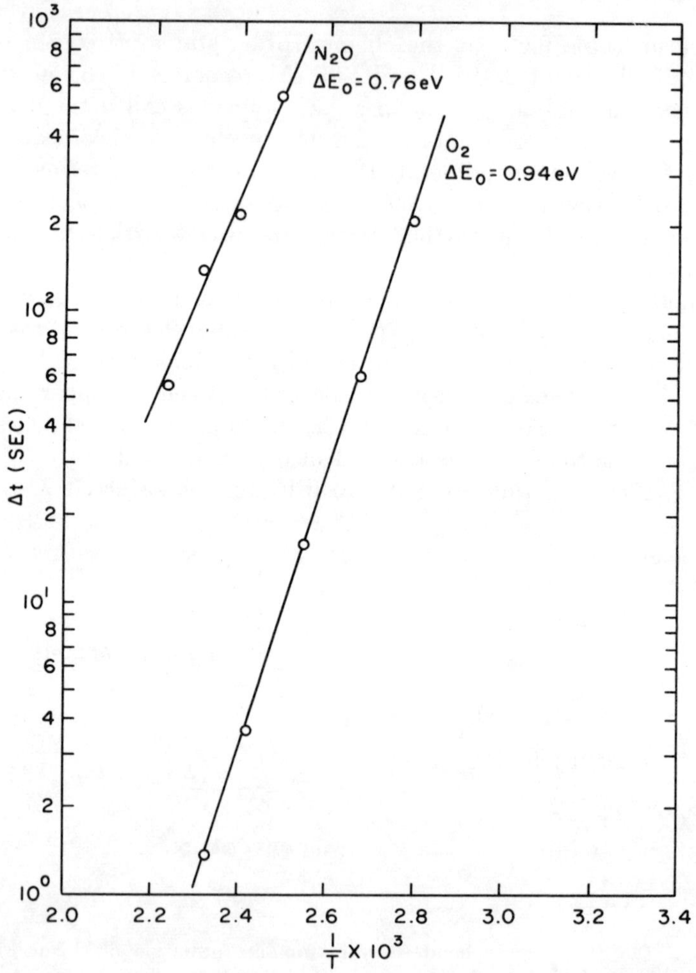

Fig. 9.5. Semilogarithmic plot of Δt versus $1/T$ for the adsorbates oxygen and nitrous oxide on the $(11\bar{2}0)$ surface of CdS (*11*). [After Mark [11]. By permission of the Pergamon Publishing Company.]

converging to their common limit, the equilibrium Fermi level. We are interested in the nonequilibrium condition that prevails during the decay period. During this period, the chemisorbed ions are desorbed on thermal release of electrons from the chemisorption level back into the conduction band. The mechanism is illustrated in Fig. 9.4b.

The rate of thermal release is

$$(dn_s/dt) = -\nu n_s \exp(-\Delta E_a/kT) \tag{18}$$

where n_s is the concentration of chemisorbed ions and ν is a frequency factor of order kT/h. Integrating Eq. (18), we obtain the chemisorbed ion concentration n_s' after a decay interval Δt,

$$\Delta t = \exp(\Delta E_a/kT)/\nu \ln(n_{s0}/n_s') \tag{19}$$

where n_{s0} is the steady-state coverage under illumination. If the solid with chemisorbed ions on its surface is reilluminated after the interval Δt, the chemisorbed ion concentration at the moment of reillumination is less than the steady-state coverage at the same illumination, $n_s' < n_{s0}$. The initial photocurrent, immediately following reillumination, is greater than the steady-state current, because the steady-state photocurrent decreases monotonically with increasing coverage of chemisorbed ions. This overshoot in photocurrent growth, which decays to the lower steady-state value as the steady-state coverage of chemisorbed ions is restored, provides the means of measuring ΔE_a. The magnitude of the overshoot $\Delta i/i_0$, where i_0 is the steady-state photocurrent, is a monotonic function of $\ln(n_{s0}/n_s')$. Finally, using Eq. (19), we obtain the decay interval required to give a particular value of the photocurrent overshoot,

$$\Delta t = t_0 \exp(\Delta E_a/kT) \tag{20}$$

where the constant t_0 is a function of $\Delta i/i_0$. Thus we obtain the desorption energy as an activation energy.

In Fig. 9.5, a typical semilogarithmic, Arrhenius-type plot, ln Δt versus $1/T$, is shown for chemisorption of oxygen and nitrous oxide on the $(11\bar{2}0)$ surface of CdS. From the slopes of the straight lines for many CdS crystals, $\Delta E_a = 0.91$ eV for oxygen and $\Delta E_a = 0.73$ eV for nitrous oxide, with an internal accuracy better than 5%.

REFERENCES

1. J. Koutecky and M. Tomasek, *Surface Sci.* **3,** 333 (1965).
2. J. Koutecky and M. Tomasek, *Int. J. Quantum Chem.* **3,** 249 (1969).
3. Th. Wolkenstein, *Advan. Catal.* **12,** 189 (1960).

4. T. B. Grimley, *Advan. Catal.* **12,** 1 (1960).
5. W. E. Garner, T. J. Gray, and F. S. Stone, *Proc. Roy. Soc. London* **A197,** 294 (1949).
6. H. M. O'Bryan and H. W. B. Skinner, *Proc. Roy. Soc.* **A176,** 229 (1940).
7. P. Mark, *J. Phys. Chem. Solids* **29,** 689 (1968).
8. J. D. Levine and P. Mark, *Phys. Rev.* **144,** 751 (1966).
9. P. Mark, *Catal. Rev.* **1,** 165 (1967).
10. J. L. Birman, *Phys. Rev.* **109,** 810 (1958).
11. P. Mark, *J. Phys. Chem. Solids* **26,** 1767 (1965).
12. R. H. Bube, "Photoconductivity of Solids," Chapter 7, pp. 233–234, Wiley, New York, 1960.
13. H. Watanabe, M. Wada, and T. Takahashi, *Jap. J. Appl. Phys.* **4,** 945 (1965).

X

The Influence of Ligand Symmetry on Chemisorption

In this chapter, we shall consider the possible arrangements of ligands—octahedral, tetrahedral, cubic, square planar, etc.—about a central metal atom and how these arrangements affect chemisorption. There are several ways of developing the picture of environmental effects: crystal-field theory, molecular orbital theory, and so-called ligand-field theory. In crystal-field theory, only electrostatic interactions between ligand and a metal ion are considered. Covalency is ignored. The ligands may be negative ions or dipoles with their negative ends directed toward the positive metal ion. We shall deal only with transition metals, because these show the most pronounced environmental effects and are most important for catalysis. In molecular orbital theory, overlapping of ligand and metal orbitals is admitted. Neither of these theories alone can explain all the observed facts, and they are often combined into an appropriate hybrid approach, termed *ligand-field theory*, which includes electrostatic effects, σ bonding, and π bonding.

Crystal-field theory is consistent with molecular orbital theory in several important points. In particular, both theories predict a splitting in the energy levels of the five degenerate d orbitals of the isolated metal atom, and the nature of the splitting depends on the structure of the ligand coordination. But the magnitude of the splitting will differ according to which theory is employed. These two theories may be considered as extremes with the true picture lying somewhere between them. As we shall see qualitatively, the nature of chemisorption is intimately associated with d-orbital splitting and the energies involved. Among chemisorption systems, we shall discuss first localized models of chemisorption on ionic solids. By a localized model is meant one which assumes the interaction of the discrete levels of a single central metal ion in the surface with the discrete orbitals of ligands in a particular coordinative structure, or one which assumes only electrostatic interactions under the same coordination. Chemisorption is pictured in this approach as completing the ligand symmetry pattern that is broken by the surface. In a second approach, electron bands in metals, especially d bands, described in terms of the overlap of molecular orbitals, may split as a result of the crystal or ligand field in the solid, thus producing directional wave functions depending on the field structure. Instead of splitting into two or more *discrete* levels, each level tends to broaden into a subband. The directions in which these bands emerge from the various crystal faces of the metal will determine how chemisorbed atoms interact.

10.1 Theories of Ligand Symmetry†

Crystal-Field Theory

The electrostatic crystal-field theory was developed by Bethe [2] and Van Vleck [3]. Consider first the influence of nearest-neighbor negative ions or oriented dipoles in octahedral coordination on the d orbitals of a transition metal ion, as shown in Fig. 10.1. The $d_{x^2-y^2}$ and d_{xy} orbitals are influenced to the same extent by ligands along the z axis, as may be seen from symmetry considerations. But ligands in the xy plane affect these orbitals differently, because the directions of maximum charge density of the $d_{x^2-y^2}$ orbital point at the ligands along the x and y axes, whereas those of the d_{xy} orbital point along the bisectors of the angles between bond directions, thus maintaining maximum separation from the ligands. Consequently, an electron in the $d_{x^2-y^2}$ orbital of the metal is repelled more strongly by the negatively charged ligands than an electron in the d_{xy} orbital. Both the $d_{x^2-y^2}$ and the d_{xy} orbitals are destabilized, but the $d_{x^2-y^2}$ orbital is destabilized to a greater extent than the d_{xy} orbital. The d_{yz} orbitals possess the same symmetry with respect to the ligands in the xz and yz planes as does the d_{xy} orbital in the xy plane. Therefore, the d_{xy}, d_{xz}, and d_{yz} orbitals must be degenerate. Just as the $d_{x^2-y^2}$ orbital

† Ref. [1].

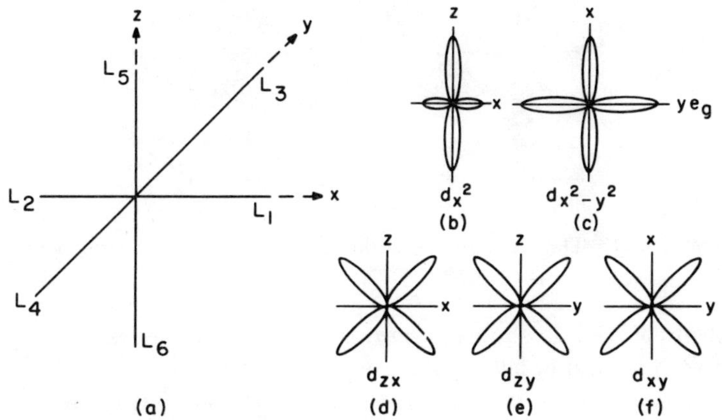

Fig. 10.1. Octahedral complex. (a) Numbering of ligands; (b)–(f) orientation of d orbitals with respect to the coordinate axes.

interacts strongly with ligands along the x and y axes, the d_{z^2} orbital interacts strongly with ligands along the z axis, and also must be unstable relative to the d_{xy}, d_{xz}, and d_{yz} orbitals. Although it is not obvious, calculations show that the $d_{x^2-y^2}$ and d_{z^2} orbitals are degenerate in an octahedral environment.

We conclude that in an octahedral environment the five d orbitals split into two groups, one group containing the three degenerate orbitals d_{xy}, d_{xz}, and d_{yz}, and the other the orbitals d_{z^2} and $d_{x^2-y^2}$. The latter group is relatively less stable, for its electrons interact more strongly with the negatively charged ligands on the x, y, and z axes. The degenerate group d_{xy}, d_{xz}, and d_{yz} is referred to as the t_{2g} orbitals and the d_{z^2} and $d_{x^2-y^2}$ group as the e_g orbitals. In Fig. 10.2a, the splitting of the d levels in an

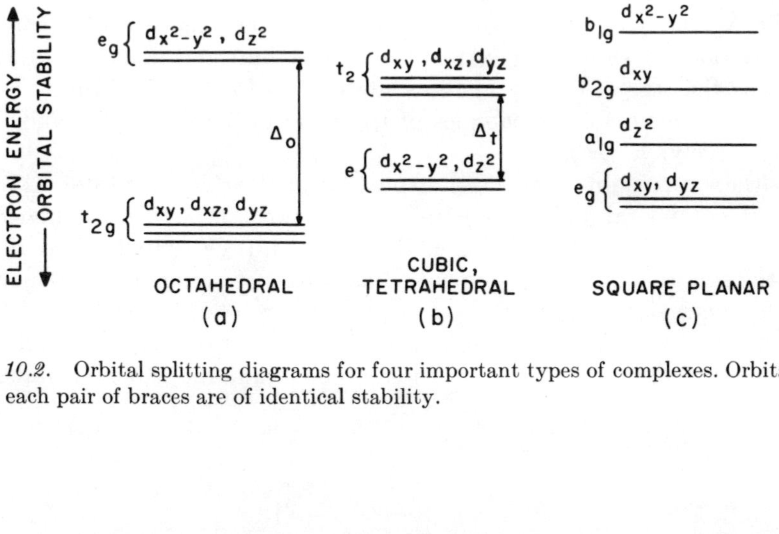

Fig. 10.2. Orbital splitting diagrams for four important types of complexes. Orbitals within each pair of braces are of identical stability.

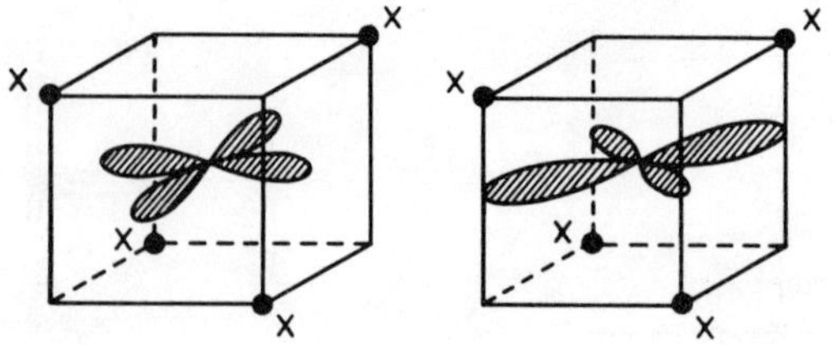

Fig. 10.3. The $d_{x^2-y^2}$ and d_{xy} orbitals in a tetrahedral environment.

octahedral complex is shown, where the symbol Δ_0 represents the energy gap between t_{2g} and e_g levels. The zero of energy is chosen as the weighted mean energy of the d orbitals, thus making the t_{2g} and e_g energies $-\frac{2}{5}\Delta_0$ and $\frac{3}{5}\Delta_0$, respectively. Such a choice is justified, for it turns out that all the useful results obtained from crystal-field calculations depend on Δ_0 and not on the absolute energies of the t_{2g} and e_g levels.

We consider next a metal ion in a tetrahedral environment. Comparing the orientation of the d_{xy} and $d_{x^2-y^2}$ orbitals relative to the ligands, we see in Fig. 10.3 that each lobe of the d_{xy} orbital comes close to one ligand and that the lobes of the $d_{x^2-y^2}$ orbital lie on the bisectors of the angles between pairs of ligands. The $d_{x^2-y^2}$ orbital should be repelled less by the ligands than the d_{xy} orbital, and calculations bear this out. Further symmetry arguments equivalent to those for the octahedral complex show that the d_{xz} and d_{yz} orbitals are degenerate with the d_{xy} orbital and that the d_{z^2} and $d_{x^2-y^2}$ orbitals are also degenerate. We see in Fig. 10.2b that the order of the split d levels is inverted relative to the octahedral environment. The levels are called t_2 and e, with the t_2 level (d_{xy}, d_{xz}, d_{yz}) being less stable. Since the orbitals are not oriented directly toward the ligands in either level, it would be expected that the splitting is smaller for tetrahedral than for octahedral complexes. Experiment and calculations confirm this. At the same distance from the metal ion, elementary calculations based on the simplest electrostatic model indicate that $\Delta_t \leq \frac{1}{2}\Delta_0$. We choose the zero of energy in the same way as for the octahedral environment. In this case, the energies of the t_2 and e levels are $\frac{2}{5}\Delta_t$ and $-\frac{3}{5}\Delta_t$, respectively.

Other important cases of the splitting of d orbitals are not as simple as the octahedral and tetrahedral coordinations, though the cubic coordination gives the same splitting as the tetrahedral (Fig. 10.2). The splitting in the square planar environment, consisting of four ligands at the corners of a square with the metal ion at the center, is difficult to assess. It is clear, however, from Fig. 10.3 that the $d_{x^2-y^2}$ orbital, whose lobes are directed toward the corners, is less stable than the other orbitals. The order of the remaining orbitals depends strongly on minor details of the model selected.

Molecular Orbital Theory†

We shall derive qualitatively the important results for octahedral complexes. It will become evident that this completely different approach gives results which are consistent in general with crystal-field theory. The LCAO–MO approximation is used and the atomic orbitals include those of the six octahedrally situated ligands and the metal ion. Initially, σ bonding will be considered exclusively, for this is the only type of bonding that can be compared between the two theories. In crystal-field theory, π bonding has no significance. The metal ion is assumed to have ns, np, and $(n - 1)$ d orbitals available for bonding. For example, in the first transition series the orbitals would be 4s, 4p, and 3d. The ligands are assumed to have only a single s orbital each. In Fig. 10.4, we illustrate the types of σ bonds which can be formed by interaction of the s orbitals of the ligands with the 4s, 4p, and 3d orbitals of the metal ion. We note that the 4s orbital of the metal interacts equally with all six ligands dis-

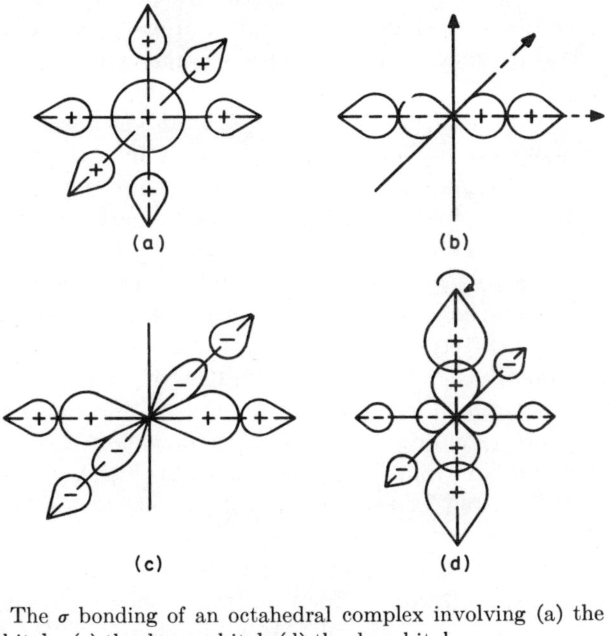

Fig. 10.4. The σ bonding of an octahedral complex involving (a) the s; (b) the p_x (p_y and p_z) orbitals; (c) the $d_{x^2-y^2}$ orbital; (d) the d_{z^2} orbital.

† Refs. [1, 4].

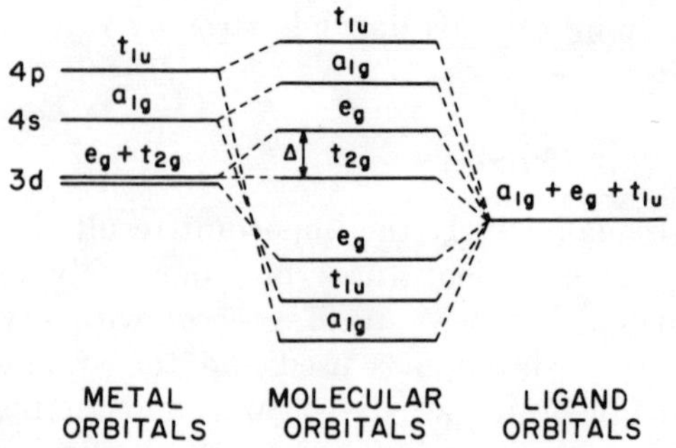

Fig. 10.5. The molecular orbital energy level diagram for an octahedral complex (schematic).

posed octahedrally (Fig. 10.4a). Interactions of p_x, p_y, and p_z metal orbitals with ligands are equivalent on the x, y, and z axes, respectively (Fig. 10.4b). The $d_{x^2-y^2}$ orbital interacts with ligands on the x and y axes (Fig. 10.4c), while the d_{z^2} orbital (Fig. 10.4d) interacts strongly with ligands on the z axis and to a lesser extent with ligands on the x and y axes. Experiment and elementary theoretical arguments indicate that interactions of $d_{x^2-y^2}$ and d_{z^2} orbitals with ligands are equivalent energetically. The d_{xy}, d_{xz}, and d_{yz} orbitals do not form σ bonds with ligands, and from symmetry considerations, are degenerate. Each interacting pair of ligand and metal orbitals forms two molecular orbitals, as discussed in earlier chapters; the pairs are designated a_{1g} for the 4s interactions, t_{1u} for the 4p, and e_g for the d_{z^2} and $d_{x^2-y^2}$, while the noninteracting d_{xy}, d_{xz}, and d_{yz} orbitals are labeled t_{2g}. In Fig. 10.5, we show schematically the disposition of the molecular orbital energy levels. The correspondence with the crystal-field picture is clear, with Δ_0 in this case giving the energy difference between the t_{2g} unperturbed level and the upper e_g level. Other environments are more difficult to interpret by LCAO–MO theory, but the solutions are similar in principle. Results for the octahedral environment are summarized in Table 10.1, including LCAO wave functions of the ligands designated ϕ_1 through ϕ_6, corresponding to the L_1–L_6 notations in Fig. 10.1a. These combinations of ligand atomic wave functions are further combined with the metal orbitals listed in the second column of the table to give the complete LCAO–MO wave function.

Wave functions for π-bonding molecular orbitals are also given in Table 10.1. We shall not go into detail for this type of bonding. An illustration of π bonding may be seen in Fig. 2.4. In addition to the d_{xy}, d_{xz}, and d_{yz} orbitals (t_{2g}), each of the three p orbitals may also be involved in π bonding. The splitting designated by Δ_0 is expected to persist, since the degenerate

TABLE 10.1

Symmetry Classifications of Orbitals for Octahedral Complexes

	Metal orbital	Ligand σ	Ligand π
a_{1g}	4s	$\psi_{a_1} = (1/\sqrt{6})(\varphi_1 + \varphi_2 + \varphi_3 + \varphi_4 + \varphi_5 + \varphi_6)$	
e_g	$3d_{z^2}$	$\psi_{eg,z^2} = (\frac{1}{2}\sqrt{3})(2\varphi_5 + 2\varphi_6 - \varphi_1 - \varphi_2 - \varphi_3 - \varphi_4)$	
	$3d_{x^2-y^2}$	$\psi_{eg,x^2-y^2} = \frac{1}{2}(\varphi_1 + \varphi_2 - \varphi_3 - \varphi_4)$	
t_{1u}	$4p_x$	$\psi_{t1u,x(\sigma)} = (1/\sqrt{2})(\varphi_1 - \varphi_2)$	$\psi_{t1u,x(\pi)} = \frac{1}{2}(\pi_{3x} + \pi_{4x} + \pi_{5x} + \pi_{6x})$
	$4p_y$	$\psi_{t1u,y(\sigma)} = (1/\sqrt{2})(\varphi_3 - \varphi_4)$	$\psi_{t1u,y(\pi)} = \frac{1}{2}(\pi_{1y} + \pi_{2y} + \pi_{5y} + \pi_{6y})$
	$4p_z$	$\psi_{t1u,z(\sigma)} = (1/\sqrt{2})(\varphi_5 - \varphi_6)$	$\psi_{t1u,z(\pi)} = \frac{1}{2}(\pi_{1z} + \pi_{2z} + \pi_{3z} + \pi_{4z})$
t_{2g}	$3d_{xy}$		$\psi_{t2g,xy} = \frac{1}{2}(\pi_{1y} - \pi_{2y} + \pi_{3x} - \pi_{4x})$
	$3d_{xz}$		$\psi_{t2g,xz} = \frac{1}{2}(\pi_{1z} - \pi_{2z} + \pi_{5x} - \pi_{6x})$
	$3d_{yz}$		$\psi_{t2g,yz} = \frac{1}{2}(\pi_{3z} - \pi_{4z} + \pi_{5y} - \pi_{6y})$

π-bonding molecular orbitals formed from the d_{xy}, d_{xz}, and d_{yz} atomic orbitals have a different energy than the σ-bonding orbitals formed from the d_{z^2} and $d_{x^2-y^2}$ atomic orbitals. In order to talk about π bonding, the simple ligand with s orbitals only must be replaced by one that possesses orbitals of π symmetry, such as an olefin. The idea of double bonding is, of course, foreign to the electrostatic crystal-field approach.

Ligand-Field Theory

Important features of the d-electron energy diagrams turn out to be equivalent in the electrostatic crystal-field theory and the molecular orbital theory, and these diagrams may be used to systematize a great deal of transition metal chemistry. But it is premature to choose between these theories. For experimental evidence shows repeatedly that both theories are needed to provide a unified theory, even in the better-known systems. A good illustration of the need for both theories lies in problems concerning the relative importance of σ-bonding, π-bonding, and electrostatic effects. As mentioned above, the π-bonding concept is utterly foreign to the electrostatic crystal-field theory. Orgel [1] and others have employed the term ligand-field theory to cover a general hybrid approach. Ligand-field splittings are considered to be the algebraic sum of a number of more or less independent mechanisms, including σ-bonding, π-bonding, and electrostatic effects. A complete, unified theory which predicts all of these

TABLE 10.2

Arrangement of Electrons and Stabilization Energies in Octahedral Complexes

Number of d electrons	Arrangements in weak ligand field: t_{2g}	Arrangements in weak ligand field: e_g	E_c (weak)[a]	Arrangement in strong ligand field: t_{2g}	Arrangement in strong ligand field: e_g	E_c (strong)[b]	ΔE_c[c]
0	—	—	0	—	—	0	0
1	↑	—	$-\frac{2}{5}\Delta_o$	↑		$-\frac{2}{5}\Delta_o$	0
2	↑ ↑	—	$-\frac{4}{5}\Delta_o$	↑ ↑	—	$-\frac{4}{5}\Delta_o$	0
3	↑ ↑ ↑	—	$-\frac{6}{5}\Delta_o$	↑ ↑ ↑	—	$-\frac{6}{5}\Delta_o$	0
4	↑ ↑ ↑	↑	$-\frac{3}{5}\Delta_o$	↑ ↓ ↑ ↑	—	$-\frac{8}{5}\Delta_o$	$-\Delta_o$
5	↑ ↑ ↑	↑ ↑	0	↑ ↓ ↑ ↓ ↑	—	$-\frac{10}{5}\Delta_o$	$-2\Delta_o$
6	↑ ↓ ↑ ↑	↑ ↑	$-\frac{2}{5}\Delta_o$	↑ ↓ ↑ ↓ ↑ ↓	—	$-\frac{12}{5}\Delta_o$	$-2\Delta_o$
7	↑ ↓ ↑ ↓ ↑	↑ ↑	$-\frac{4}{5}\Delta_o$	↑ ↓ ↑ ↓ ↑ ↓	↑	$-\frac{9}{5}\Delta_o$	$-\Delta_o$
8	↑ ↓ ↑ ↓ ↑ ↓	↑ ↑	$-\frac{6}{5}\Delta_o$	↑ ↓ ↑ ↓ ↑ ↓	↑ ↑	$-\frac{6}{5}\Delta_o$	0
9	↑ ↓ ↑ ↓ ↑ ↓	↑ ↓ ↑	$-\frac{3}{5}\Delta$	↑ ↓ ↑ ↓ ↑ ↓	↑ ↓ ↑	$-\frac{3}{5}\Delta_o$	0
10	↑ ↓ ↑ ↓ ↑ ↓	↑ ↓ ↑ ↓	0	↑ ↓ ↑ ↓ ↑ ↓	↑ ↓ ↑ ↓	0	0

[a] Weak-field stabilization energy.
[b] Strong-field stabilization energy.
[c] E_c(strong) $-$ E_c(weak) $= \Delta E_c$.

mechanisms as limiting cases from basic quantum-mechanical concepts is, of course, desirable, but not yet available.

Stabilization Energies

When the energy gap Δ for any coordinative arrangement is small, electrons will fill the five d orbitals singly before any pairing of electrons occurs, as in the case of the free ion. But when Δ is large, degenerate orbitals of the lower level are assumed to fill completely with a pair of electrons in each orbital before filling of the orbitals of the upper level commences. In Table 10.2, we show the arrangement of electrons in the d orbitals in a weak field (Δ small) and a strong field (Δ large) for the octahedral and tetrahedral coordinations.

An important quantity is the stabilization energy E_c referred to as the crystal-field or ligand-field stabilization energy (CFSE or LFSE). This quantity is based on the energy assignments given earlier: $-\frac{2}{5}\Delta_0$ for the t_{2g} level and $\frac{3}{5}\Delta_0$ for the e_g level in octahedral coordination and $-\frac{3}{5}\Delta_t$ for the e level and $\frac{2}{5}\Delta_t$ for the t_2 level in tetrahedral coordination. The stabili-

TABLE 10.3

Stabilization Energies in Octahedral and Tetrahedral Fields[a]

Number of electrons	Stabilization energy[b] E_c (octahedral)	E_c (tetrahedral)	ΔE_c [c]
0, 5, 10	0	0	
1, 6	$-\frac{2}{5}\Delta_o$	$-\frac{3}{5}\Delta_t$	$-\frac{1}{10}\Delta_o$
2, 7	$-\frac{4}{5}\Delta_o$	$-\frac{6}{5}\Delta_t$	$-\frac{2}{10}\Delta_o$
3, 8	$-\frac{6}{5}\Delta_o$	$-\frac{4}{5}\Delta_t$	$-\frac{8}{10}\Delta_o$
4, 9	$-\frac{3}{5}\Delta_o$	$-\frac{2}{5}\Delta_t$	$-\frac{4}{10}\Delta_o$

[a] Weak-field approximation.

[b] Δ_t is generally $\frac{1}{2}\Delta_o$ for a given ion with common ligands.

[c] E_c(octahedral) − E_c(tetrahedral) with $\Delta_t = \frac{1}{2}\Delta_o$, illustrating the twin-peaked pattern.

zation energies are expressed by

$$E_c(\text{octahedral}) = -\tfrac{2}{5}n_{t_{2g}}\Delta_0 + \tfrac{3}{5}n_{e_g}\Delta_0 \tag{1}$$

and

$$E_c(\text{tetrahedral}) = -\tfrac{3}{5}n_{t_2}\Delta_t + \tfrac{2}{5}n_e\Delta_t, \tag{2}$$

where $n_{t_{2g}}$ and n_{e_g} are the number of electrons in the t_{2g} and e_g levels of the octahedral environment and n_{t_2} and n_e the number in the t_2 and e levels of the tetrahedral environment. Each of these equations is used for both strong- and weak-field arrangements, with proper attention to the order of filling the orbitals in the two cases. These energies are given in Table 10.2 for octahedral coordination, strong- and weak-field arrangements, and in Table 10.3 for octahedral and tetrahedral coordination, weak-field arrangement only. In the last column of Table 10.2, the quantity ΔE_c shows that the strong-field arrangement is either equally stable or more stable than the weak-field arrangement, depending on the number of d electrons, and the last column of Table 10.3 shows that octahedral coordination is either as stable as or more stable than tetrahedral coordination, again depending on the number of d electrons.

Stabilization energies and their differences are functions of the magnitude of the energy separation Δ. We give the gross trends in magnitude of Δ_0

set forth by Orgel [1] as obtained from electronic adsorption spectra in octahedrally coordinated compounds:

1. The complexes of a given ligand with ions of the same row of the periodic table in the same valency have Δ_0 values in a fairly narrow range. Thus Δ_0 for the hydrates of divalent ions of the first series varies from 7800 cm^{-1} for Mn^{2+} to about 11,000 cm^{-1} for the Cr^{2+} ion.

2. For the complexes of a given ligand, Δ_0 increases rather rapidly with the valency of the metal ion. Thus, while the hydrates of divalent ions have $\Delta_0 \approx 10{,}000$ cm^{-1}, the Δ_0 values for trivalent ions of the first series cluster about $\Delta_0 \approx 20{,}000$ cm^{-1}.

3. In the corresponding complexes of metals of different transition series, Δ_0 increases, often about 30%, from the first transition series to the second, and by a similar amount for the second to the third series.

4. The common ligands can be arranged in an order of increasing ligand-field strength such that Δ_0 increases along the series in a manner more or less independent of the particular metal ion studied. Thus the relative size of Δ_0 is a characteristic of a ligand.

10.2 Localized Models of Chemisorption on Ionic Solids

Applications of ligand-field theory to chemisorption are not yet well developed; often they consist merely of pictorial representations of molecular orbitals, geometrical diagrams of charged ligands about a central metal ion, or empirical correlations based on meager data. Dowden and Wells [5, 6] were among the first to suggest that the phenomena of chemisorption and catalysis could be explained through considerations of localized symmetry. They pointed out that rate-controlling steps in certain chemisorptions and catalytic reactions which involve species so polarized as to restore near octahedral symmetry about a surface cation in an ionic solid contribute to higher energies of activation and lower activities at the cation electron configurations d^0, d^5, and d^{10}. Such configurations possess zero stabilization energies. In most transitions from some configuration to the octahedral configuration, the change in stabilization energy (ΔE_c) leads to an increase in stability of the complex ($\Delta E_c < 0$). Thus maxima in ΔE_c frequently appear between d^0 and d^5 and between d^5 and d^{10}, as shown in Table 10.3 for the transition from tetrahedral to octahedral symmetry. A rough similarity to the twin-peaked ΔE_c pattern is found for the activity of hydrogen reactions as a function of the number of electrons in the d band in the first transition series, and this suggests that ΔE_c may be a significant factor in the energetics of chemisorption and catalysis. Similar correlations hold for oxygen chemisorption.

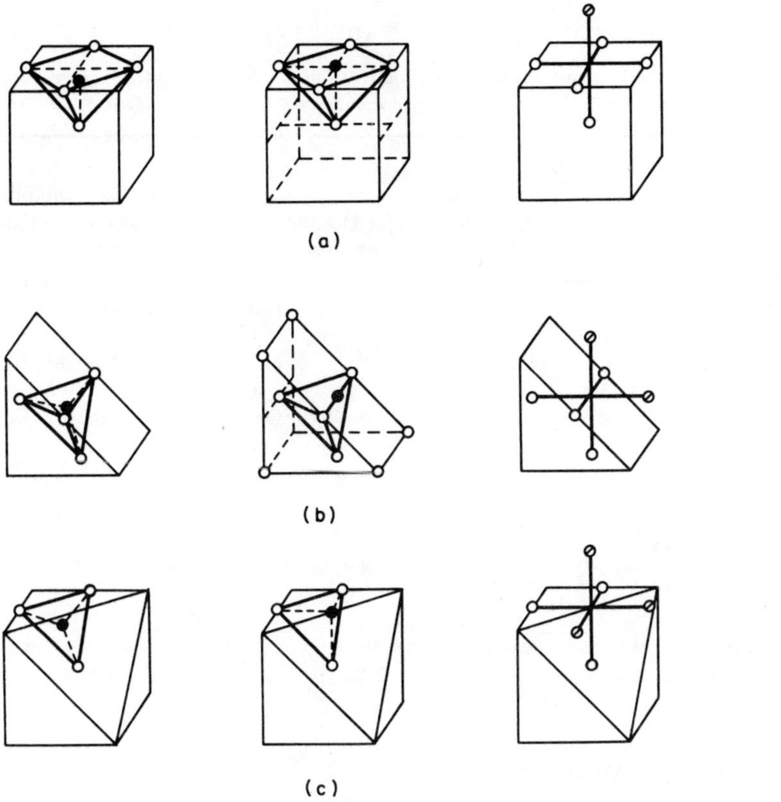

Fig. 10.6. Changes in nickel ion coordination during the chemisorption of oxygen on nickel oxide: (a) (100) plane; (b) (110) plane; (c) (111) plane. Diagrams in the middle column refer to a bare surface at the instant of cleavage; left column, bare surface, relaxed position; right column, after adsorption of oxygen, ● nickel ion; ○ lattice oxygen ions; ⊖ adsorbed oxygen ions (7). [After Haber and Stone [7]. By permission of The Faraday Society.]

Haber and Stone [7] applied the general observations of Dowden and Wells to the specific case of chemisorption of oxygen on nickel oxide. They considered the chemisorption of oxygen anions (O^{2-}) on the three principal faces (100), (110), and (111) illustrated in Figs. 10.6a–c, respectively. The surface as it appears at the moment of cleavage is represented in the middle column; the bare surface after relaxation is shown in the left column, and the surface after chemisorption of oxygen is seen in the right column. Sufficient adsorption on each face to give octahedral symmetry is assumed to take place. A nickel ion in the (100) face (Fig. 10.6a) has five nearest-neighbor oxygen ions distributed in the shape of a square pyramid. After

TABLE 10.4

Changes in Crystal-Field Stabilization Energy during Chemisorption of Oxygen on Nickel Oxide[a]

Plane	Field	Stabilization energy [b]	Atoms adsorbed	Final field	Stabilization energy [b]	Gain in stabilization energy	Gain per adsorbed atom [c]
(100)	Square pyramid	$-10.0Dq$	1	Octahedral	$-12Dq$	$2Dq$	$2Dq$
(110)	Tetrahedral	$-3.6Dq$	2	Octahedral	$-12Dq$	$8.4Dq$	$4.2Dq$
(111)	Trigonal	$-10.9Dq$	3	Octahedral	$-12Dq$	$1.1Dq$	$0.4Dq$

[a] See Haber and Stone [7].
[b] $Dq = 910\ \text{cm}^{-1} = 2.6$ kcal mole^{-1}.
[c] $10Dq = \Delta_o$, where Δ_o is the d-level separation energy in an octahedral field.

cleavage, the surface may relax in such a way that the nickel ion enters the pyramid, as shown to the left in Fig. 10.6a. Actually, oxygen ions move relative to the nickel ion, but for simplicity only movement of the nickel ion is depicted. Chemisorption of one oxygen ion completes the octahedron about the Ni^{2+} ion, as shown to the right in Fig. 10.6a. To estimate the change in stabilization energy in the transition from square pyramidal to octahedral coordination, the data given by Basolo and Pearson [8] are used. For d^8 octahedral symmetry, they give a stabilization energy of $-12Dq$, where $10Dq = \Delta_0$ (Table 10.3) is another nomenclature for the d-level separation energy in an octahedral field. For the stabilization energy in square pyramidal d^8 symmetry, they give $-10\ Dq$ $(= -\Delta_0)$. So the gain in stabilization energy is $2Dq$ $[= \frac{1}{5}\Delta_0]$ per adsorbed atom on the (100) face. Similar changes occur on the (110) and (111) faces, as shown in Figs. 10.6b and c, respectively. The relaxed position for the (110) face approximates tetrahedral symmetry for the cation and, upon adsorption of *two* oxygen ions, octahedral symmetry is attained, with a gain in stabilization energy of $8.4Dq$. A nickel ion on a freshly cleaved (111) face relaxes toward the center of the triangle of oxygen ions lying directly beneath it, giving trigonal symmetry. Octahedral symmetry is attained by adsorption of *three* oxygen ions. The gain in stabilization energy is $1.1Dq$. Results are summarized in Table 10.4, from which it is evident that the theory predicts the greatest stabilization for the

(110) face. At low overall coverage, we may expect (110) faces to be preferentially covered at the expense of the (100) and (111) faces.

To support their theory, Haber and Stone [7] carried out photodesorption experiments. They found that adsorbed oxygen on nearly stoichiometric nickel oxide could be desorbed by illumination, the most effective region being the visible spectrum in the wavelength range 650–900 mμ. Using an Orgel diagram [9] for the Ni^{2+} ion in tetrahedral and octahedral fields, which gives the energy levels for these symmetries, Haber and Stone concluded that oxygen photodesorption occurs by a transition from octahedrally coordinated nickel in its ground state $^3A_{2g}$ to the excited state $^3T_{1g}$—a transition whose adsorption spectra appear around 650 mμ. The excited state $^3T_{1g}$, they believe, is an activated complex unstable with respect to the ground state of a tetrahedrally coordinated Ni^{2+} ion. Thus a surface Ni^{2+} ion which has chemisorbed two oxygen atoms to form an octahedral complex may lose the two atoms as an oxygen molecule and return to the tetrahedral state when irradiated with the wavelength corresponding to the transition $^3A_{2g} \rightarrow {}^3T_{1g}$. Jongepier and Schuit [10] check the experimental results of Haber and Stone, but have serious doubts concerning their interpretation in spite of its plausibility. Among the most serious criticisms of Jongepier and Schuit are (1) no allowance is made for charge transfer and valence changes of the nickel ion on desorption of neutral oxygen; (2) it is not possible to have a continuous transition from O_h to T_d symmetry while the $^3T_{1g}$ wave function conserves its symmetry. They give no alternative interpretation, stating that a sufficiently comprehensive theoretical model based on molecular orbital treatment is not yet available.

In esr studies, Shvets and Kazansky [11] have investigated oxygen anion radicals adsorbed on supported oxide catalysts containing Ti, V, and Mo ions. They find that the valence state of the transition metal ions and their reducing properties are primarily responsible for electron transfer from the adsorption center to the adsorbed molecule. For example, in a series of isoelectronic ions with the configuration $3d^1$ (Ti^{3+}, V^{4+}, and Cr^{5+}), the oxygen molecule anion O_2^- is formed only for Ti and V. No O_2^- is found on Cr^{5+} because of its oxidizing properties. The anion O^- is stable on Ti^{3+} only at low temperatures, for Ti^{3+} gives off electrons at high temperatures (>300°C) which reduce the O^- anion radical to diamagnetic lattice O^{2-}. On supported V^{4+}, O^- is thermally stable on account of the poor reducing properties of V^{4+} ions.

Another factor which they find affects the nature of adsorbed oxygen is the state of coordination of the transition metal surface ion. This coordination is affected by the crystalline structure of the support. For example,

Ti, V, and Mo ions are in tetrahedral coordination when supported on SiO_2 and show great differences in oxygen anion adsorption characteristics from the unsupported oxides. In such systems, where both valency and coordination affect adsorption characteristics, fundamental theoretical interpretations are extremely difficult.

Crystal-field effects are associated not only with the metal ions of the adsorbent, but also with the chemisorbed particles, that is, the adsorbate induces a field in the metal and the metal induces a field in the adsorbate. Kollrack [12, 13] has calculated stabilization energies for hydrogen, nitrogen, oxygen, sulfur, fluorine, and chlorine adsorbed on third-row transition metal oxide catalysts. Under certain geometric arrangements, the stabilization energy for O^- may be as high as 70 kcal $mole^{-1}$ [12]. In general, the stabilization energies may be ordered as follows: $S > O \approx Cl > F$, with no stabilization of N and H at all [13].

10.3 Field Effects and the Band Structure of Metals

Electron bands have been described in terms of the overlap of orbitals which are directional because of the crystal field of the metal [14, 15]. The approach is an extension of the ligand-field theory employed in the study of complex molecules [1, 4, 16]. Trost [14] has worked out the orbital construction and electron population for bands in transition metals.

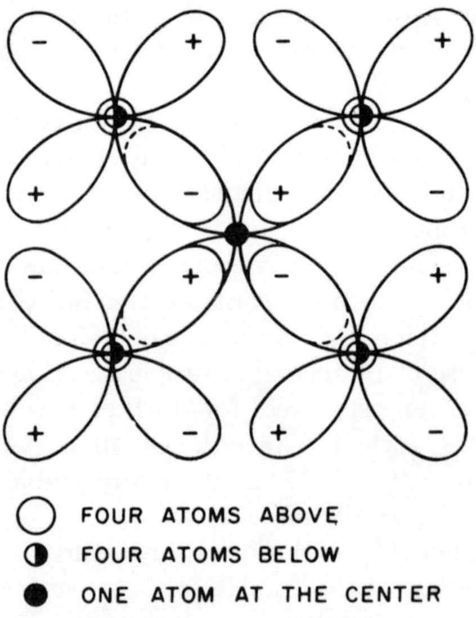

Fig. 10.7. Body-centered dϵ (t_{2g}) band; four atoms above, four atoms below.

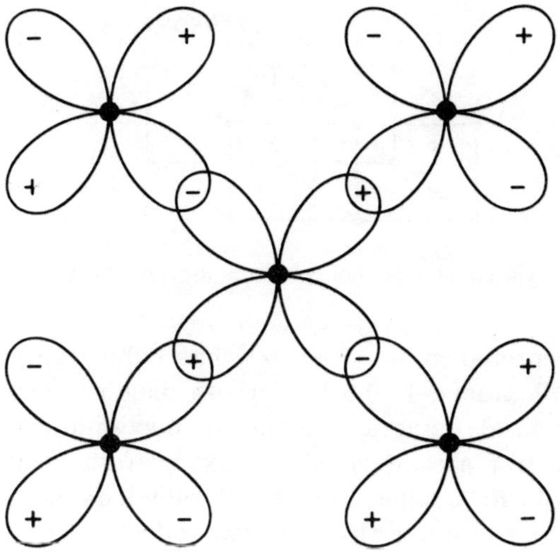

Fig. 10.8. Cubic close-packed dϵ; five atoms in plane.

Assuming uniform density of states in delocalized bonds among near neighbors, he has been able to predict correctly the type of lattice for all transition metals except Mn and Fe. Considering the three degenerate p orbitals, he shows how they overlap to give a body-centered structure. The resulting bcc p band has cubic symmetry, is perfectly degenerate in 3 p orbitals, is delocalized over all atoms, and is a close-packed assembly of p orbitals. But the corresponding case for d orbitals shows that not all five orbitals are degenerate, unless hybridization occurs, because all d orbitals cannot achieve the same overlap at the same time in any one lattice. As in the case of molecular complexes, the orbitals separate into diagonal (d_{xy}, d_{yz}, and d_{xz}) and axial (d_{z^2} and $d_{x^2-y^2}$). There are two close-packed structures possible for the diagonal orbitals. In the first, one orbital overlaps eight others, as shown in Fig. 10.7. The d_{xy}, d_{yz}, and d_{xz} orbitals are all equivalent, with eight orbitals overlapping in each case; degeneracy is preserved, and a body-centered lattice established. The band is delocalized over all atoms in the crystal for these orbitals and is called the bcc dϵ band. Frequently it is called the t_{2g} band, for it corresponds to the t_{2g} orbitals of molecular complexes discussed earlier. In the second way, four orbitals overlap with a central one, as shown in Fig. 10.8. When the three diagonal orbitals (d_{xy}, d_{xz}, d_{yz}) all overlap in this way, 12 atoms surround the central one in cubic close-packed arrangement. Again, the three orbitals are equivalent, and the cubic close-packed dϵ band (ccp dϵ),

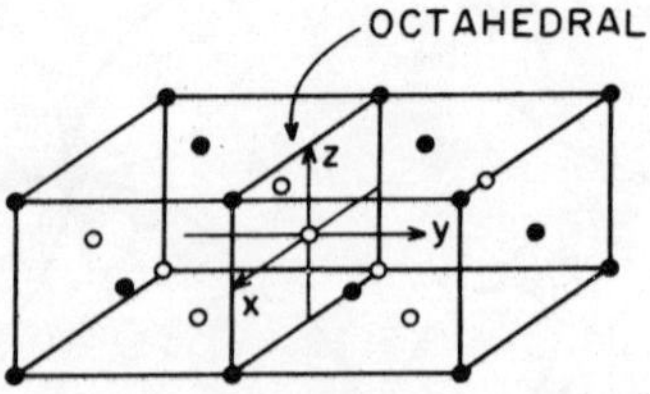

Fig. 10.9. Cartesian frame of reference for face-centered cubic metal.

like the body-centered cubic (bcc dϵ), has cubic symmetry and is delocalized over all atoms. In both of these diagonal orbital lattices, the $d_{x^2-y^2}$ and d_{z^2} orbitals have a direction of maximum extension identical with the octahedral arrangement of next-nearest neighbors. They are degenerate and form a delocalized band called dγ or e_g. The threefold degenerate dϵ band is lower than the twofold degenerate dγ band. With one valence electron, we find that electron confined to the s and p bands. As the number of electrons increases, the energy levels of the d orbitals fall relative to the s and p. With five electrons, the d orbitals are dominant, holding approximately three of the five electrons.

We discuss a similar picture for face-centered cubic nickel given by Goodenough [15]. In Fig. 10.9, the face-centered cubic structure is con-

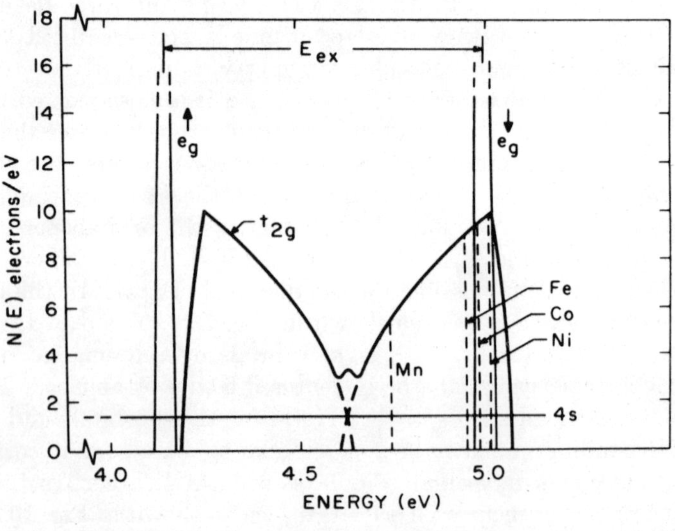

Fig. 10.10. Schematic density of states curve for face-centered cubic nickel. Fermi levels are given for Ni, Mn, Fe, and Co. [After Goodenough [15]. By permission of John Wiley and Sons.]

sidered with a Cartesian frame of reference. The orientation of the dϵ (t_{2g}) electron bands and the dγ (e_g) electron bands are shown in Fig. 10.1b–f. There are 12 nearest-neighbor atoms and the central atom may be bonded to each of them as a result of each of the 12 lobes of its dϵ (t_{2g}) orbitals (four in each of the three planes) overlapping with the lobe of a nearest neighbor. There are six next-nearest-neighbor atoms, and each of the six lobes of the dγ (e_g) orbitals of the central atom is directed toward one of them along the coordinate axes, but not necessarily bonded to them because of the greater distance of separation. In Fig. 10.10, the density of states curve, $N(E)$ versus E, is given for nickel, including the crystal-field modification of the bands. We see that the dγ (e_g) band is narrow, indicating that the orbitals are nearly localized and that they do not overlap neighboring orbitals significantly. The nearly localized dγ (e_g) electrons occupy two levels (narrow bands) resulting from intraatomic exchange splitting (Hund's rule). On the contrary, the dϵ (t_{2g}) band is broadened appreciably by orbital overlapping, though it is still narrow compared to the s band. From polarized neutron experiments, Goodenough estimates that the 0.55 hole of the d band is split into 0.41 t_{2g} and 0.14 e_g hole.

10.4 Nonlocalized Model of Chemisorption

On the basis of the split d-band structure of transition metals, Bond [17] has introduced a qualitative theory of chemisorption. Referring to Figs. 10.1 and 10.9, we can describe the emergence of orbitals on the various faces of a face-centered cubic metal. For example, two dϵ (t_{2g}) orbitals emerge at 45° (d_{yz}, d_{xz}) on the (100) face; one dϵ (t_{2g}) orbital, the d_{xy}, is parallel to the (100) surface and does not emerge (0°); one dγ (e_g) orbital, the d_{z^2}, emerges normal to the (100) surface (90°), and one dγ (e_g) orbital, the $d_{x^2-y^2}$, is parallel and does not emerge. Emergence on the (110) and (111) faces may be determined similarly, and the information is summarized in Table 10.5. Only the (111) face has no vertically emerging orbitals.

Turning to chemisorption, we recall that the interaction of a nonlocalized level of the metal band structure with a discrete level of an adsorbate does not lead necessarily to a nonlocalized chemisorptive bond. It is possible, depending on the values of the quantum-mechanical parameters, for the bond to be localized outside the range of band energies. By a nonlocalized model, we mean in this case a model that involves interaction with the d-band structure of the metal as opposed to the localized ligand-field model of Section 10.2, which considers only the central metal atom and its nearest-neighbor ligands.

TABLE 10.5

Angle of Emergence of Orbitals in the Crystal Faces of a Face-Centered Cubic Metal

Subband	Orbital	Face (100)	(110)	(111)
$d\epsilon$ (t_{2g})	d_{xy}	0°	30° (2)	30° (1)
$d\epsilon$ (t_{2g})	d_{xz}	45° (2) [a]	30° (2)	30° (1)
$d\epsilon$ (t_{2g})	d_{yz}	45° (2)	90° (1)	30° (1)
$d\gamma$ (e_g)	$d_{x^2-y^2}$	0°	45° (1)	36°16′ (2)
$d\gamma$ (e_g)	d_{z^2}	90° (1)	45° (1)	36°16′ (1)

[a] Numbers in parentheses are numbers of lobes that emerge.

A possible site for adsorption of a hydrogen atom on a (100) surface is the vertically emerging $d\gamma$ (e_g) orbital, which could overlap with the 1s orbital of hydrogen. An alternative site is the position marked "octahedral site" in Fig. 10.9, where a 1s orbital could be overlapped by five $d\gamma$ (e_g) orbitals, giving a strongly bound state.

Monoolefins should adsorb most readily on the (100) face of nickel using the partially vacant e_g (90°) and the t_{2g} (45°) orbitals to form a π-adsorbed complex (see Fig. 2.4). A coordinate bond could form on the (110) face between the olefin and an e_g orbital with the bond axis at 45° to the surface. If the t_{2g} and e_g orbitals can interchange functions, a bond similar to that on the (100) face is conceivable. Similarly, there are linear sequences e_g, t_{2g}, e_g on the (111) face which could serve for π adsorption of ethylene, provided the functions e_g and t_{2g} orbitals can interchange. There are no linear sequences t_{2g}, e_g, t_{2g} on the (111) face, and the e_g orbitals on this face make an angle of only about 36° with the surface, which could lead to considerable repulsion between the adsorbed molecule and neighboring surface atoms. Two possibilities arise for σ-diadsorbed olefins: (1) overlap with adjacent vertical e_g orbitals on the (100) face; and (2) overlap with adjacent vertical t_{2g} orbitals or with 45°-emergent e_g orbitals on the (110) face. Again, the (111) face appears to be the least suited for chemisorption of olefins. Further examples are given by Bond of the qualitative application of the ligand-field approach to chemisorption of carbon monoxide, diolefins, and alkynes with reasonable results.

Though Bond's results are reasonable in their qualitative state, the question of their validity for the adsorption of olefins on specific transition metals is not yet answerable. Jansen [18] sounds a dubious note. He points out that such adsorption requires strong crystal-field splitting. But accurate neutron diffraction measurements on several transition metals [19, 20] have shown that the d-electron orbitals are spherical to a good approximation, indicating that crystal-field splitting is relatively weak.

10.5 Summary

It will now have become obvious that many theoretical predictions about the chemisorptive bond cannot be checked experimentally because of limitations in methods available for examining surfaces. Also, many experimental facts about real surfaces cannot be derived theoretically because of the indicated complexity of the model. We conclude with a few brief remarks on the limitations.

Many techniques and instruments have been applied to the study of the chemisorptive bond. The following list, though not complete, gives the more important ones.

Kinetic studies
Semiconductivity measurements
Surface photoillumination
Adsorption isotherms
Flash desorption
Work function measurements
Calorimetry
Ellipsometry
Mass spectrometer
Electron diffraction
Mass spectrograph
Thermally stimulated electron current measurements
Radial electron distribution
Magnetic susceptibility
Photoelectric measurements
Low-energy electron diffraction (LEED)
X ray
Auger spectroscopy
Electron spectroscopy for chemical analysis (ESCA)
Mossbauer spectroscopy
Field emission microscopy
Electron probe microanalysis
Scanning electron microscope
Low-energy back-scattered ions
Sputter ion source instruments
Diffuse reflectance spectrometry
X-ray fluorescence spectrometry
Phosphorescence spectroscopy
Thermoluminescence measurements

In spite of innumerable attacks by this formidable array of weapons, the fortress of chemisorption continues to guard successfully many secrets. For instance, only in a few special cases has it been possible to determine the number of chemisorption sites upon which catalytic reaction occurs. And frequently these cases seem to describe a dismal picture of site concentrations so low (10^7–10^{12}/cm^2) that even the most sophisticated and modern techniques are helpless. Out of the welter of adsorption sites on a given surface, differing in both degree and kind, no clear picture of the nature of the catalytically active ones has yet emerged. We have pointed out repeatedly the continuing battle of localized versus nonlocalized chemisorption. Theory sets forth the parameters by which the two states are distinguishable. But as yet the parameters are not measurable.

We are speaking about the ultimate experiments that we should like to perform. Without a knowledge of catalytically active site concentrations, kinetic studies to elucidate mechanisms are hampered, for they can be expressed only in terms whose relations to site concentrations may be arbitrary; and undoubtedly the catalytic activity of an adsorbed molecule is reflected by the degree of localization of the chemisorptive bond.

From the point of view of theory, we are equally handicapped. In spite of theoretical considerations that distinguish metals from ionic lattices, all models of chemisorption depict idealized surfaces which may be periodic or aperiodic, but always smooth and regular. It is difficult to visualize how the irregularities revealed by experiment, such as reconstruction, defects, and impurities, could be handled effectively. Moreover, the general nature and obscure localized geometry of models of adsorbed molecules give a diffuse caricature of reality.

We cannot be optimistic about solving these fundamental limitations. But short of such ultimate goals, there are many important victories to be won through the cooperation of theory and experiment.

REFERENCES

1. L. E. Orgel, "An Introduction to Transition-Metal Chemistry, Ligand-Field Theory." Methuen, London, 1960.
2. H. Bethe, *Ann. Phys.* **3,** 133 (1929).
3. J. H. Van Vleck, "Theory of Electric and Magnetic Susceptibilities." Oxford Univ. Press, London and New York, 1932.
4. J. S. Griffith, "Theory of Transition Metal Ions." Cambridge Univ. Press, London and New York, 1961.
5. D. A. Dowden and D. Wells, *Actes 2e Congr. Int. Catal.* p. 1499. Technip, Paris (1961).
6. D. A. Dowden, *Catal. Rev.* **5,** 1 (1971).
7. J. Haber and F. S. Stone, *Trans. Faraday Soc.* **59,** 192 (1963).
8. F. Basolo and R. G. Pearson, "Mechanisms of Inorganic Reactions." Wiley, New York 1958.
9. L. E. Orgel, *J. Chem. Phys.* **23,** 1004 (1955).
10. R. Jongepier and G. C. A. Schuit, *J. Catal.* **3,** 464 (1964).
11. V. A. Shvets and V. B. Kazansky, *J. Catal.* **25,** 123 (1972).
12. R. Kollrack, *J. Catal.* **12,** 321 (1968).

13. R. Kollrack, *J. Catal.* **14,** 332 (1969).
14. W. R. Trost, *Can. J. Chem.* **37,** 460 (1959).
15. J. B. Goodenough, "Magnetism and Chemical Bond." Wiley (Interscience), New York, 1963.
16. C. J. Ballhausen, "Introduction to Ligand-Field Theory." McGraw-Hill, New York, 1962.
17. G. C. Bond, *Discuss. Faraday Soc.* **41,** 200 (1966).
18. L. Jansen, *Mol. Proc. Solid Surfaces, Battelle Inst. Mater. Sci. Colloq. 3rd* p. 49 (1968).
19. R. Nathans, G. C. Shull, G. Shirane, and A. Andersen, *Phys. Chem. Solids* **10,** 138 (1959).
20. N. N. Sirota and N. Olekhnovich, *Sov. Phys. Dokl.* **6,** 704 (1962).

A 4
B 5
C 6
D 7
E 8
F 9
G 0
H 1
I 2
J 3

Author Index

Numbers in parentheses are reference numbers and indicate that an author's work is referred to although his name is not cited in the text. Numbers in italics show the page on which the complete reference is listed.

Subject Index

L

M

N

O

P

R

S

Physical Chemistry

A Series of Monographs

Ernest M. Loebl, Editor

Department of Chemistry

Polytechnic Institute of New York

Brooklyn, New York

1 W. JOST: Diffusion in Solids, Liquids, Gases, 1952
2 S. MIZUSHIMA: Structure of Molecules and Internal Rotation, 1954
3 H. H. G. JELLINEK: Degradation of Vinyl Polymers, 1955
4 M. E. L. MCBAIN and E. HUTCHINSON: Solubilization and Related Phenomena, 1955
5 C. H. BAMFORD, A. ELLIOTT, and W. E. HANBY: Synthetic Polypeptides, 1956
6 GEORGE J. JANZ: Thermodynamic Properties of Organic Compounds — Estimation Methods, Principles and Practice, Revised Edition, 1967
7 G. K. T. CONN and D. G. AVERY: Infrared Methods, 1960
8 C. B. MONK: Electrolytic Dissociation, 1961
9 P. LEIGHTON: Photochemistry of Air Pollution, 1961
10 P. J. HOLMES: Electrochemistry of Semiconductors, 1962
11 H. FUJITA: The Mathematical Theory of Sedimentation Analysis, 1962
12 K. SHINODA, T. NAKAGAWA, B. TAMAMUSHI, and T. ISEMURA: Colloidal Surfactants, 1963
13 J. E. WOLLRAB: Rotational Spectra and Molecular Structure, 1967
14 A. NELSON WRIGHT and C. A. WINKLER: Active Nitrogen, 1968
15 R. B. ANDERSON: Experimental Methods in Catalytic Research, 1968
16 MILTON KERKER: The Scattering of Light and Other Electromagnetic Radiation, 1969
17 OLEG V. KRYLOV: Catalysis by Nonmetals — Rules for Catalyst Selection, 1970
18 ALFRED CLARK: The Theory of Adsorption and Catalysis, 1970

19 ARNOLD REISMAN: Phase Equilibria: Basic Principles, Applications, Experimental Techniques, 1970

20 J. J. BIKERMAN: Physical Surfaces, 1970

21 R. T. SANDERSON: Chemical Bonds and Bond Energy, 1970

22 S. PETRUCCI, ED.: Ionic Interactions: From Dilute Solutions to Fused Salts (In Two Volumes), 1971

23 A. B. F. DUNCAN: Rydberg Series in Atoms and Molecules, 1971

24 J. R. ANDERSON: Chemisorption and Reactions on Metallic Films, 1971

25 E. A. MOELWYN-HUGHES: Chemical Statics and Kinetics of Solution, 1971

26 IVAN DRAGANIC AND ZORICA DRAGANIC: The Radiation Chemistry of Water, 1971

27 M. B. HUGLIN: Light Scattering from Polymer Solutions, 1972

28 M. J. BLANDAMER: Introduction to Chemical Ultrasonics, 1973

29 A. I. KITAIGORODSKY: Molecular Crystals and Molecules, 1973

30 WENDELL FORST: Theory of Unimolecular Reactions, 1973

31 JERRY GOODISMAN: Diatomic Interaction Potential Theory. Volume 1, Fundamentals, 1973; Volume 2, Applications, 1973

32 ALFRED CLARK: The Chemisorptive Bond: Basic Concepts, 1974

In Preparation

I. G. KAPLAN: Symmetry of Many-Electron Systems

SAUL T. EPSTEIN: The Variation Method in Quantum Chemistry